# STUDY GUIDE
# AND
# PSYCH JOURNAL

## Brian Cole
## George Semb
*University of Kansas*

Fifth Edition

# PSYCHOLOGY
## Principles and Applications

## Stephen Worchel
## Wayne Shebilske

**PRENTICE HALL,** *Englewood Cliffs, New Jersey 07632*

©1995 by PRENTICE-HALL, INC.
A Simon and Schuster Company
Englewood Cliffs, New Jersey 07632

10 9 8 7 6 5 4 3 2 1

ISBN 0-13-563578-0
Printed in the United States of America

# CONTENTS

**Preface**  iv

    To the Student   iv
    How to Answer Multiple-Choice Questions   iv
    To the Instructor   v
    About the Authors   vi

**Study Guide Chapters**

# PREFACE

This **Study Guide and Psych Journal** is designed to help you learn the material in <u>Psychology</u>, Fifth Edition. The **Study Guide** portion consists of a set of questions or learning objectives that will promote and reinforce your understanding of the topics covered in the text. The **Psych Journal** entries at the end of each chapter will also help involve you in learning more about psychology and how it applies to you.

Each chapter in the study guide begins with an introduction or **Chapter Summary**. This is followed by a list of **Key Terms and Concepts**. Next is a series of **Short-Answer Discussion Exercises**, followed by a **Posttest** consisting of a series of true-false, multiple-choice, and fill-in-the-blank questions that will reinforce what you have learned. After you complete the test, you can check your answers against those at the end of the chapter. Each chapter concludes with **Psych Journal** questions drawn directly from the text.

We recommend that you begin your study of each chapter by scanning it, noting main headings, and carefully reading the entire assignment. Next, read the main points in this study guide and define the key terms and concepts and respond to the short-answer questions in your own words. If a concept or question is confusing or if you do not fully understand it, then reread the appropriate section of the text. If you then find yourself copying a response directly from the text, it probably means that your understanding of the concept is still unclear. In that case you should discuss the material with another student or the instructor.

We have designed this study guide to encourage your active participation in the learning process. This is **the** most important variable in determining how well you will learn. Do the exercises and you **will be** successful. We guarantee it.

## INSTRUCTIONS FOR ANSWERING MULTIPLE-CHOICE QUESTIONS

There are several things you should know about how to answer the multiple-choice questions that appear on the **POSTTEST** at the end of each chapter.

First, each item has *one* and only *one* correct answer. Second, if a question has an alternative that states "all of the above," "all of the above are correct," or "both x and y are correct," you should select the *most comprehensive* alterative. For example, consider the following items:

1.    Sacramento is:
    a.    a state capital.
    b.    located in California.
    c.    a city with a population of more than 150,000 people.
    d.    all of the above.

Because alternatives a, b, and c are all correct, d is the best answer.

2.  Which of the following statements is correct?
    a.  What you become is a function of both heredity and the environment.
    b.  The concept of "average" is useful in a statistical sense, but not representative of the individual.
    c.  Both a and b are correct.
    d.  Neither a nor b is correct.

Because both a and b are accurate statements, c is the best alternative.

3.  Which of the following statements is correct?
    a.  Sigmund Freud was the father of cognitive psychology.
    b.  B. F. Skinner constructed the first test of intelligence (IQ).
    c.  Both a and b are correct.
    d.  Neither a nor b is correct.

Because both a and b are inaccurate statements, d is the best alternative.

4.  Which of the following statements is correct?
    a.  Most parents underestimate and understate the abilities of their children, particularly when talking with close friends and relatives.
    b.  The most dramatic motor development a child makes during infancy is learning to walk.
    c.  Most children do not show evidence of depth perception until they reach middle childhood.
    d.  All of the above are correct.

Because a and c are inaccurate statements, b is the correct answer.


## TO THE INSTRUCTOR

Psychology, Fifth Edition, and this study guide lend themselves equally well to lecture classes, individualized instruction, modular scheduling, computer-managed instruction, the Personalized System of Instruction (PSI), and small group discussions. This study guide is suitable for students who need step-by-step instruction or who simply wish to review the material quickly. Its aim is to help students learn the material you assign.

The structure of each chapter involves your students in the learning process, particularly the sections headed **Key Terms and Concepts** and **Discussion Questions and Exercises**. Questions in Test Item File for Psychology, Fifth Edition, build on the material outlined in the study guide. In addition, the test items contain several questions for each key term and concept to permit repeated testing and several challenging exercises that require students to analyze, evaluate, integrate, and apply the material they have learned.

If you have any questions please write to Brian Cole or George Semb, Department of Human Development, University of Kansas, Lawrence, Kansas, 66045.

## ABOUT THE AUTHORS

Brian Cole is a graduate student in instructional design in the Department of Human Development at the University of Kansas. He has taught child development courses for the past several years. He specializes in instructional design, assessment of structural knowledge, long-term retention, and tutor training.

George Semb is a professor of instructional design at the University of Kansas. He has written study guides, instructor's manuals, and computer software for texts in psychology, aviation, behavior analysis, family studies, and biology. His primary area of interest is in the design, delivery, and evaluation of instructional systems. He has written several articles dealing with instructional design, individualized instruction, peer tutoring, and on-the-job training. Many of the techniques he has investigated are incorporated in this study guide.

# Chapter 1 -- The World of Psychology

**CHAPTER SUMMARY**

1.      **Psychology** is the scientific study of behavior and application gained from that knowledge. The challenge to the science of psychology is to identify the most likely situations under which certain behaviors will occur. The application of psychology involves knowing which principle to use, and how and when to apply it.

2.      The **field of psychology** has developed through stages characterized by division and consolidations. Psychology has always had a close relationship with the social issues of the time, and these issues have often helped identify new areas of focus. Psychologists as well as their subjects have been predominantly white and male. Psychologists are trying to change this situation by welcoming more minority members to the field and by increasing their awareness of the influence of culture and behavior.

3.      **Structuralism** focused on identifying the elements of human experience and finding out how these elements interact to form thoughts and feelings. **Functionalists** were concerned with the function of mental processes to fill needs. This emphasis led to important applications in education and the founding of educational psychology. **Gestalt psychologists** insist that our experiences or behaviors must be analyzed as wholes; they should not be broken down into elements. According to Freud, the founder of **psychoanalytic psychology**, human behavior is the result of internal conflicts involving largely unconscious motivations. Many, if not most, of these thoughts and wishes are the result of our experiences during infancy and early childhood. The **humanistic approach** is based on the idea that people are basically good, and, given the proper environment, they will strive to achieve positive social goals. **Behaviorists** believe that human behavior can be explained and predicted by focusing on only observable stimuli and responses. According to the **cognitive approach**, people process, evaluate, and interpret information and events, and their responses are shaped as much by these "subjective realities" as by the physical reality of the events themselves.

4.      **Physiological psychology** studies the biological basis of behavior. **Cognitive psychology** has its roots in experimental psychology and is concerned with the mental activities involved in acquiring, processing, and storing information. **Developmental psychologists** examine the function of age on behavior. **Social psychology** is the study of the way people are affected by social situations. **Personality psychologists** focus on explaining and predicting the unique ways that people respond to their environment. **Clinical psychologists** diagnose and treat emotional and behavioral disorders; they include health psychologists and community psychologists. **Counseling and school psychology** concentrate on helping people with social, educational, and career adjustments. **Other subfields** include engineering psychology, industrial-organizational psychology, environmental psychology, forensic psychology, psychology of minorities, and law and psychology.

5.      Students with a bachelor's degree in psychology find themselves well received in law schools, medical schools, and business schools. Other settings that welcome graduates with degrees in psychology include hospitals, mental health clinics, businesses, and government agencies.

6.    Psychologists develop hypotheses about how people will behave in certain situations. Based on theories, hypotheses are tested using different research methods. The case history involves studying a few individuals (sometimes only one) or the effects of a single event in depth. To collect data from a large population, psychologists use surveys or questionnaires. Through naturalistic observation, psychologists observe people in their own settings. This method often leads to discovering correlations. In an experiment, researchers directly study variables to determine which events actually cause certain behaviors. The independent variable is manipulated. The dependent variable is measured to see the effects on it of the manipulation of the independent variable. Randomization, the use of placebos, and double-blind control insure that the results of the experiment are unbiased. Ethical issues concerning the rights of human and animal subjects are important to researchers. These issues include invasion of privacy, deception, possible harmful consequences, and humane treatment of animals.

7.    Psychologists use statistical analysis to interpret and communicate the results of research. Using statistical techniques, psychologists determine the probability of certain behaviors occurring by chance. Statistics also help psychologists evaluate the relationship between two variables as in a correlation.


## KEY TERMS AND CONCEPTS

What is Psychology?
    The scientific study of behavior and applications gained from that knowledge

History and Scope of Psychology
    Plato asked "Why are people different?"
    Descartes was concerned with the concept of "mind"
    Field of psychology was created in late 1800s
        Early psychologists sought understanding through observation
        Developed through stages of division and consolidation
        The field is large

Increasing Psychology's Scope: Culture, Race, and Gender
    Although roots are European, the field is strong in United States
    Until recently psychology was dominated by white men
    Last two decades, the field has made strides toward race sensitivity
        There is an increasing awareness of "Culture" on behavior
        Individualism vs. Collectivism

Perspectives in Psychology
    Structuralism - identifying the elements of human experience, how they interact
        Developed by Wundt and Titchener
            Wundt started the first psychological laboratory
        Analytic introspection - isolate the elements of which experience is made
            The elements: sensations, feelings, and images
    Functionalism - emphasizes the function of thought to fill needs
        Proposed by William James
        Led to important applications in education and educational psychology
    Gestalt Psychology - based on premise that we experience whole sensations
        Established by Kohler, Koffka, Wertheimer
        "Gestalt" - German word meaning "whole"

Psychoanalytic Psychology - treatment of anxiety disorders by dealing with
     repressed feelings
     Introduced by Sigmund Freud
          Freud put the European scientific world in an uproar
          His ideas had a tremendous influence on psychology
Humanistic Approach - states that people are basically good
     This view rejects the Freudian and behaviorist view of people
     Popular in clinical psychology
Behaviorism - behavior is described only on observable stimuli and response
     John Watson was one of the first behaviorists
     Well-known behaviorist - B. F. Skinner
Cognitive Approach - we process, evaluate, and interpret information

Subfields of Psychology - the many field of psychology
     Physiological Psychology - learning, memory, perception studied by
          neurobiological events
     Experimental Psychology - behaviors and cognitions related to learning
     Cognitive Psychology - roots are in traditional experimental psychology
     Developmental Psychology - function of age on behavior
     Social Psychology - study of way people are affected by social relations
     Personality Psychology - study of how a person adjusts to the environment
     Clinical, Community, and Health Psychology - treatment of emotional disorders
     Counseling and School Psychology - helps people make adjustments in careers
     Engineering and Industrial-Organizational Psychology - contact with tools and
          machines as comfortable and error free as possible
     Other Subfields

Careers in Psychology
     Psychology is one of the most popular majors in America
     Psychologists have many job options
          Employed by hospitals, mental health clinics, businesses

Research Methods in Psychology
     Basic Principles - hypotheses (guesses) and theories (explanations)
     Case History - in depth look at a few individuals or effects of a single event
          Advantage - easy method in a short time within natural setting
          Disadvantage - uses only a few cases
     Survey - uses questionnaires given to large samples of people
          Advantage - collect a great deal of information, flexibility
          Disadvantage - memory in earlier settings not always clear
     Naturalistic Observation - people's reactions to naturally occurring events
          Advantage - firsthand observations
          Disadvantage - results difficult to verify, no cause-effect relation
               Correlation - measure of the extent to which variables change
               together
     Experiment - a researcher directly manipulates one variable to another variable
          Independent variable - manipulated variable
          Dependent variable - measured variable
          Ways to ensure subject classification and avoid bias
               Randomization - random assignment to groups
               Placebos - used in drug studies, made of inactive material
               Double-blind control - used to ensure there is no experiment bias

Strengths of experiments
       They establish cause-effect relationships
       They can be repeated by anyone to verify them
       Used to analyze precise variables because they can be
          manipulated
Weaknesses of experiments
       Subjects know they are being studied
       Sometimes independent and dependent variables unrealistic
       Limited subject populations

Ethical Issues in Research
       No research should be undertaken unless the investigator determines that the issue being studied is important for advancing our understanding of the behavior
       There are guidelines published by the APA
          Invasion of privacy
          Deception
          Harmful consequences

Looking at the Numbers: Statistical Analysis - communication of results of a study
       Help to determine the properties of data


## DISCUSSION QUESTIONS AND EXERCISES

Note: These questions and exercises ARE the learning objectives for this chapter. Answer them accurately in your own words and you will have mastered the most important material. We guarantee it.

1.   <u>What is Psychology?</u>

a.   Define **psychology**? What are the goals of science?

*Study of human behaviour, scientific study of.*

*Goals describe, predict & explain events.*

b.   According to Kelley, what is scientific psychology and it's purpose?

*is the knowledge gained from through the use of scientific method.*

*gives us sense of order to our knowledge, help eliminate confusion.*

4

c. The understanding of psychology requires: _learning theory's, results; research, process used to obtain that knowledge._

2. <u>History</u> <u>and</u> <u>Scope</u> <u>of</u> <u>Psychology</u>

a. If we examine books on the history of psychology what would we find? _that many begin with a discussion of the greeks, espec Plato._

b. When was the field of psychology created? With what event was it created?

c. Characterize the stages of psychology as it developed through the following time periods:

   (1) Late 1800s and early 1900s

   (2) Early 1900s until 1960

(3)    1990s

d.    Where were the roots of psychology?  Where is it most strongly embraced?

e.    What has the increasing awareness of the influence of culture on behavior provided us with?  What is culture?

3.    Perspectives in Psychology

a.    Who started the first psychological laboratory?  What techniques did he use?

b.    Describe **structuralism** and the main research tool utilized in structuralism.

c.   Who proposed **functionalism**? Define functionalism and give an example of what you might study as a functionalist.

d.   What does the word **"Gestalt"** mean in German, and characterize Gestalt psychology?

e.   Who introduced the **psychoanalytic approach**?   What are the basic principles of psychoanalysis?

f.   What influence has Freud's ideas had on psychology, and why?

g. Give a brief description of the following three approaches:

(1) **Humanistic approach**  *free will*

(2) **Behaviorism**  *observable behavior*

(3) **Cognitive approach**  - *all things in consideration*

4. <u>Subfields</u> <u>of</u> <u>Psychology</u>

a. Match each of the following subfields of psychology with its appropriate description:

1. Cognitive Psychology _____ study of biological basis of behavior
2. Experimental Psychology _____ mental activities in storing information
3. Social Psychology _____ study of issues such as learning, sensation
4. Developmental Psychology _____ study of individual differences
5. Physiological Psychology _____ way people are affected by social situations
6. Personality Psychology _____ examines the function of age on behavior
7. Clinical Psychology _____ contact with tools error free as possible
8. Counseling Psychology _____ helps with job or social adjustments

b. Identify three other subfields of psychology and characterize each one.

5. <u>Careers</u> <u>in</u> <u>Psychology</u>

a. Why is psychology such a popular major?

b. What job options are there for psychology majors?  Identify some of the settings in which people with study in this area can be useful.

6. <u>Research</u> <u>Methods</u> <u>in</u> <u>Psychology</u>

a. Trace the procedures of a psychologists doing research from beginning to end.

b. Define **hypotheses** and **theory**.

theory - set of principaly that
explain why behavour ocurs

hypotheses - predictions in a measurable.

c.      What method is used to look in depth at a few individual cases?  Identify at least two of its advantages and at least two of it disadvantages.

d.      What is one way around the disadvantage of focusing only on a few people in the **case history** method?  Identity some problems with the use of **surveys**.

e.      Characterize **naturalistic observation**.  Identify at least two advantages of using this method of observation and at least two disadvantages.

f.        What is a **correlation**? Explain in your own words. Give at least one original example of a correlation.

g.        Identify the components of an **experiment**. Present your own example of an experiment identifying all the components.

h.        What is one method an experimenter would ensure each subject in an experiment had an equal chance of being assigned to any group? Why is this important? Explain and give an example.

i.    How can an experimenter avoid the bias that subject expectations may bring to an experiment?  Give an example.

j.    How can a researcher counteract the effects of experimenter bias?  Give an example that illustrates how this counteractive measure might work.

k.    Identify at least three strengths and three weaknesses of using an experiment to demonstrate the existence of a cause and effect relationship.  Provide an example in which you analyze the pros and cons of experimentation.

7. <u>Ethical</u> <u>Issues</u>

a. Why is it important to not overlook the issue of ethics in research?

b. Identify some of the ethical principles provided to you in your book regarding research with humans in psychology.

8. <u>Looking</u> <u>at</u> <u>the</u> <u>Numbers:</u> <u>Statistical</u> <u>Analysis</u>

a. What is the purpose of statistical analyses?

b. What is the general rule among psychologists on the existence of **"real"** differences in psychological research?

## POSTTEST

1.  From the following statements about psychology, which one is NOT correct?
    a.  Both animals and human beings are studied in psychology
    b.  Psychology studies development throughout the lifespan, from conception till death
    c.  Psychology takes a scientific approach to understanding behavior
    d.  Psychology uses philosophical methods to answer questions such as "What is the meaning of life?"

2.  One of the goals of science is to _____ behavior.
    a.  direct
    b.  explain
    c.  motivate
    d.  teach

3.  The philosopher/psychologist Descartes' position that humans behaved according to laws and mechanistic principles would, today, be most closely aligned with which school of psychology?
    a.  behavioral
    b.  cognitive
    c.  humanistic
    d.  structural

4.  What is the challenge to the science of psychology?
    a.  to develop a wide range of approaches, applications, and methods
    b.  to develop an understanding of everyday concerns
    c.  to identify specific causes of behavior
    d.  to identify the most likely situations under which certain behaviors will occur

5.  Which of the following does NOT fit with the other three?
    a.  functionalism
    b.  introspection
    c.  Titchener
    d.  Wundt

6.  Modern psychology is thought to have begun when:
    a.  Freud developed psychoanalytic theory and therapy
    b.  Greek philosophers began studying the differences between people
    c.  it was discovered that many illnesses have no medical basis
    d.  Wundt opened the first laboratory devoted to the scientific study of psychology

7.  Plato believed that:
    a.  birth determines our differences
    b.  in many hidden ways people are equal
    c.  individual differences are determined at birth
    d.  people are similar

8. Descartes was concerned with the _concept of mind_.
   a. behavior of animals
   b. concept of mind
   c. laws and principles
   d. rules and standards

9. The roots of psychology are:
   a. American
   b. East Asian
   c. European
   d. French

10. Which early approach in psychology focused on breaking conscious experiences into their fundamental components?
    a. behaviorism
    b. functionalism
    c. humanism
    d. structuralism

11. Imagine yourself standing in front of a window looking at a house, trees, and a sky. Next, divide the window scene into different colors of brightness and shade. According to structuralists, what are you doing?
    a. analytic introspection
    b. analytic psychology
    c. deductive reasoning
    d. inductive reasoning

12. William James proposed a different program of research that emphasizes the function of thought. This program is called _____a_____.
    a. functionalism
    b. gestalt psychology
    c. psychoanalysis
    d. structuralism

13. If you were asking the question "What is thinking for?" you would be a:
    a. behaviorist
    b. functionalist
    c. psychoanalyst
    d. structuralist

14. Gestalt is a German word meaning ___d___.
    a. element
    b. part
    c. sum
    d. whole

15. Freud put the European scientific world in an uproar with his unorthodox views of human behavior. Freud was the founder of _____d_____.
    a. behaviorism
    b. functionalism
    c. psychosocialism
    d. psychoanalysis

16. Which approach to human behavior rejects the Freudian view of people?
    a.  behavioral
    b.  cognitive
    c.  humanistic
    d.  psychoanalytic

17. {T  F} Humanistic psychology believes that humans are basically good.

18. {T  F} Behaviorism argued that if psychology was to become a science, it must focus on replicable events.

19. Henry conducts research on the effects of reward and punishment on maze learning in mice.  Henry is most likely a:
    a.  behaviorist
    b.  cognitive psychologists
    c.  functionalist
    d.  structionalist

20. A well-known advocate of behaviorism was:
    a.  B. F. Skinner
    b.  Carl Rodgers
    c.  Sigmund Freud
    d.  William James

21. Which of the following is NOT likely to be of concern to a cognitive psychologist?
    a.  how we store things in memory
    b.  the role of the consequences of an animal's behavior on the frequency of that behavior
    c.  why it is easier to remember some things but not others
    d.  why we pay attention to some things but not others

22. Which of the following psychologists would study the neurological events that occur during learning?
    a.  clinical
    b.  developmental
    c.  experimental
    d.  physiological

23. Sommers, in a series of studies conducted in the 1970s, showed that furniture arrangement and room decorations were directly related to how nursing home clients interacted with one another.  Dr. Sommers would, today, be classified as a(n) _____ psychologist.
    a.  community
    b.  environmental
    c.  forensic
    d.  health

24. Which of the following types of psychologists is most prepared to help a person who hears voices and believes she is Joan of Arc?
    a.  clinical
    b.  counseling
    c.  forensic
    d.  school

25. A researcher, based on his review of relevant scientific studies, believes that there is a relationship between the frequency of a baby's crying and whether it was nursed at set intervals or on a demand schedule.  The researchers belief is best described as a:
   a. fact
   b. hypothesis
   c. prediction
   d. theory

26. A set of principles that explains why behaviors occur is termed a:
   a. concept
   b. hypothesis
   c. theory
   d. variables

27. When two factors are allowed to vary in a scientific investigation, these factors are called:
   a. cases
   b. instances
   c. theories
   d. variables

28. A major problem with the case study method is that it:
   a. doesn't provide a rich source of research hypothesis
   b. is useless when one wants to study rare events
   c. is performed in artificial, contrived settings
   d. lacks generality

29. A weakness of _____ is that people want to be "good" subjects and give the experimenter what they want.
   a. case histories
   b. laboratory experiments
   c. questionnaires
   d. surveys

30. As part of a class assignment, Hilary's professor asked the members of the class to complete anonymous questionnaire on prejudice.  Which research method was Hilary's professor using?
   a. field experiment
   d. laboratory experiment
   c. naturalistic observation
   b. survey

31. {T  F} In naturalistic observation, a drawback is that the results are difficult to verify.

32. A study is conducted to see if a protein-enriched diet will enhance the maze-running performance of rats. One group of rats is fed the high-protein diet during the study; the other group continues to receive the ordinary rat meal. In this experiment, the group of rats that is fed the high-protein diet is a(n) _____ group; the group that receives ordinary rat chow is a(n) _____ group.
    a. control; control
    b. control; experimental
    c. experimental; control
    d. experimental; experimental

33. If two variables increase or decrease together at the same time they are:
    a. inversely correlated
    b. negatively correlated
    c. positively correlated
    d. uncorrelated

34. {T  F} The existence of a correlation almost always indicates a cause-and-effect relationship.

35. A researcher found that clients who were randomly assigned to same-sex groups participated more in group therapy sessions than clients who were randomly assigned to coed groups. In this experiment, the independent variable was:
    a. how much the clients' mental health improved
    b. the amount of participation in the group therapy sessions
    c. the clients' attitudes toward group therapy
    d. whether or not the group was coed

36. Police officers are given varying amounts of alcohol, then tested for driving ability. The result of the driving ability test is the _____.
    a. control variable
    b. dependent variable
    c. independent variable
    d. sample variable

37. What is the purpose of randomization?
    a. to equate subjects
    b. to insure equal compliance
    c. to maintain the standards of the experiment
    d. to train subjects

38. The expectation on the part of an experimenter that a subject will behave in a certain way is known as called experimenter _____.

39. In an effort to insure that neither the subject nor the experimenter is aware of how the independent variable is being manipulated, which of the following was procedures was developed?
    a. double-blind control
    b. randomization
    c. stratified sampling
    d. variable control

40. Dr. Smith is going to do an experiment with drugs. He is concerned that his research subjects will respond to demand characteristics. He may want to control this by using which of the following?
   a. a placebo
   b. control groups
   c. randomization
   d. two independent variables

41. {T F} No research project should be undertaken unless the investigator determines that the issue being studied is important for advancing our understanding of behavior.

42. In an effort to protect the subject's right to privacy in research, which of the following policies is employed?
   a. deception, so that subjects will not be influenced by the real/underlying reason for the study
   b. informed consent
   c. obtaining permission from a parent, spouse, or guardian if the person is unable or unwilling to provide their own permission
   d. requiring the person to sign a release form that protects not only the human rights of the subject, but also those of the experimenter

43. Psychologists have generally adopted the rule that if the results of a study could have happened by chance less than 5 percent of the time a _____ occurs.
   a. "ethical" difference
   b. "real" difference
   c. "statistical" difference
   d. "variable" difference

44. Statistical analysis of data allows one to determine which of the following?
   a. whether demand characteristics were eliminated
   b. whether experimenter bias could have produced the obtained results
   c. whether the study was ethical
   d. whether there is a relationship between two variables

45. {T F} If someone has volunteered to be a subject in a psychological experiment, it is no longer necessary to keep the data on them confidential.

46. Maria describes herself as friendly, outgoing, and intelligent. Based on this information, in what kind of culture did Maria probably grow up?
   a. collectivist
   b. communitarian
   c. democratic
   d. individualistic

47. Rolonda and Karen want to do research on aggression. Rolonda is interested in trying to determine what aggression is by isolating the various feelings and sensations that are involved in aggressive behavior. Karen is more interested in discovering what purpose is served by aggressive behavior. It seems that Rolonda is taking a _____ approach whereas Karen is taking a _____ approach.
   a. functionalist, Gestalt
   b. functionalist, structuralist
   c. Gestalt, functionalist
   d. structuralist, functionalist

48. One important difference between the psychoanalytic and behaviorist approaches is that the psychoanalytic approach focuses on _____ whereas the behaviorist approach focuses on _____.
   a. elements of experience, unconscious impulses
   b. observable behavior, unconscious impulses
   c. unconscious impulses, elements of experience
   d. unconscious impulses, observable behavior

49. Clinical psychologists _____ whereas counseling psychologists _____.
   a. diagnose and treat emotional and behavioral disorders; help people with social, educational, or career adjustments
   b. examine how a particular life-style affects one's physical health; focus on changed in our physical and social environment
   c. focus on changes in our physical and social environment; examine how a particular life-style affects one's physical health
   d. help people with social, educational, or career adjustments; diagnose and treat emotional and behavioral disorders

50. One advantage of being able to control the variables in a research experiment is:
   a. it eliminates the possibility of experimenter bias
   b. it eliminates the possibility of demand characteristics
   c. it ensures that a cause and effect relationship will be found
   d. variables can be analyzed precisely

## POSTTEST ANSWERS

1. d
2. b
3. a
4. d
5. d
6. a
7. c
8. b
9. c
10. d
11. a
12. a
13. b
14. d
15. d
16. c
17. T
18. F
19. a
20. a
21. b
22. d
23. b
24. a
25. b
26. c
27. d
28. d
29. d
30. b
31. T
32. c
33. c
34. F
35. d
36. b
37. a
38. bias
39. a
40. a
41. T
42. b
43. b
44. d
45. F
46. d
47. d
48. d
49. a
50. d

# PSYCH JOURNAL

**Please use the following pages to record your thoughts and feelings about the following questions.**

1.  Compare and contrast how issues of culture, race, and gender in the field of psychology are reflected in your life and Robeson's life.

_____

_____

_____

_____

_____

_____

_____

_____

_____

_____

_____

_____

_____

_____

_____

_____

_____

_____

_____

_____

_____

2. Review the studying guide in Chapter 1. How do these suggestions differ from your usual studying habits? What methods do you think you'll adopt?

_____

_____

_____

_____

_____

_____

_____

_____

_____

_____

_____

_____

_____

_____

_____

_____

_____

_____

_____

3. Have you or has anyone you know ever gone to a therapist? What approaches did the therapist use?

_____

_____

_____

_____

_____

_____

_____

_____

_____

_____

_____

_____

_____

_____

_____

_____

_____

_____

4. Have you ever felt that your viewpoint is not represented by a group you belong to? What viewpoint would you like to add to this "white male European" canon in psychology?

_____

_____

_____

_____

_____

_____

_____

_____

_____

_____

_____

_____

_____

_____

_____

_____

_____

_____

_____

_____

5. Has anyone ever betrayed your confidence about a vary private matter?  Why is confidentiality so important in psychology?

_____

_____

_____

_____

_____

_____

_____

_____

_____

_____

_____

_____

_____

_____

_____

_____

_____

_____

_____

_____

_____

# Chapter 2 -- Biology and Behavior

## CHAPTER SUMMARY

1.      **Biopsychology** has historical roots in early Greek philosophies.  At the turn of the nineteenth century, Franz Gall introduced **phrenology**, focusing attention on specific areas and functions of the brain by drawing a map of the brain's bumps and hollows.  Modern **biopsychologists**, using brain imaging techniques to study the function and structure of the brain, have located networks of control centers.  Other biological entities including the **nervous system**, the **endocrine system**, and **genetics** play a part in behavior.

2.      The brain contains between 100 and 200 billion neurons.  Neurons have tiny branch-like fibers (dendrites) that receive messages from surrounding neurons; a single long extension called an **axon** carries messages away from the cell body toward other neurons.  Myelin sheaths composed of glial cells cover the axons and allow a neural impulse to move quickly.  **Synaptic transmission** occurs as a neural impulse moves from one neuron to another.  An incoming impulse must be above a minimum level or threshold of intensity for a cell to fire.  A neuron sends different messages, not by changing the strength of its response but by changing the rate of its response.  During synaptic transmission, a chemical released by one neuron travels to another.  All **neurotransmitters** carry messages between neurons, but they each carry specific messages.  **Acetylcholine** is involved in complex behaviors.  **Catecholamines** have been implicated in mental disorders.  Research on **endorphins** has yielded insights into alcoholism and other addictive behaviors.

3.      The peripheral nervous system consists of the somatic nervous system and the autonomic nervous system.  The **somatic nervous system** carries information from the eyes, ears and other senses to the brain.  It also carries information from the brain to the muscles.  The **autonomic nervous system** has two divisions: the **sympathetic division** (especially busy during stressful situations) prepares us for action; the **parasympathetic division** allows us to relax.

4.      The **spinal cord** connects the peripheral nervous system to the brain and also controls reflexes.  The brain's anatomical subdivisions include the hindbrain, midbrain, and forebrain.  The **hindbrain** includes the medulla, pons, and cerebellum.  The **midbrain** contains the reticular formation, a critical mass of neurons involved with alertness and consciousness.  The **forebrain** includes the hypothalamus, thalamus, and cerebrum.  The **cerebrum** governs our most advanced human capabilities including abstract reasoning and speech.  The cerebrum is divided into left and right brain hemispheres.  Each hemisphere predominantly services the opposite side of the body.  Research suggests that the **left hemisphere** understands abstract language, enables people to speak, and performs complicated mathematical computations.  The **right hemisphere** is superior in drawing, assembling blocks to match designs, and recognizing emotions in facial expressions.  The **reticular activating** system controls waking and sleeping.  The **limbic system** plays an important role in goal-directed behavior, memories, and emotions.  The **motor system** controls voluntary muscle actions.  Damage to the cortex may affect speech and cause **Broca's** or **Wernicke's aphasia.**

5.      The glands of the **endocrine system** release hormones (chemical messengers) directly into the bloodstream, and the bloodstream carries the hormones to other body tissues. **Growth hormones** affect our height. **Sex hormones** regulate our reproductive abilities and secondary sex characteristics such as the distribution of body hair and breast development. Metabolism is controlled by the **thyroid gland**. **Adrenal glands** secrete hormones that govern the flight-or-fight response. The **pancreas** affects energy levels by controlling the level of sugar in the blood.

6.      **Genetics** is the study of how traits are inherited, or passed on, from parent to child. **Genes**, cells located in the nucleus of every cell in the body are composed of **deoxyribonucleic acid (DNA)**, which contains the blueprints for life. Genes are grouped together in chromosomes. Every human should have 23 chromosome pairs. Most human traits are **polygenic**, which means they are determined by the action of more than one gene pair. Therefore, it is difficult to understand which genes affect behavior. Errors in the transmission of genetic information results in mutations, such as **Klinefelter** and **Down syndrome**. **Geneticists** are tracking down genetic connections with behavior and how environment influences heredity.

## KEY TERMS AND CONCEPTS

History and Scope of Biopsychology
     Roots traced to Socrates and Plato
     Joseph Gall proposed theory of phrenology
         Bumps in the head are significant
         This theory was incorrect
     Focus is on the brain and nervous system
     Today theories are being developed about biological determinants of behavior

Neurons: Basic units of the nervous system
     Brain's basic components
         Neuron cells
         Glial cells (glia - glue) - blood-brain barrier
             Microglia - cleaning system
         Brain contains 100 to 200 billion neurons
             Dendrites serve as connections
             Axons carry messages away from cell bodies
             Myelin sheath is the fatty covering of axons
     Neural impulses
         Resting potential - the "off" state
             Polarization - resting state of tension
         Acting potential - the "on" state
             Depolarization
     Synaptic transmission - occurs as an impulse moves from neuron to neuron
         Synapse - junction between two neurons
         Five parts of a synapse
             Axon terminal
             Dendrite of receiving cell
             Synaptic vesicles
             Synaptic space
             Receptor sites
         Absolute refractory period - once a cell fires, it can't fire again for 1 ms.
         Relative refractory period - a cell only fires for a strong impulse

Chemical transmitters - neurotransmitters (50 identified)
    Acetylcholine - most common
    Catecholamines - instability in these lead to mental disorders
    Dopamine
    Norepinephrine
    Serotonin

Peripheral nervous system - consists of all parts of nerve cells that lie outside brain
    Run to and from organs, glands, muscles, and sensors
    Somatic nervous system - information from senses to the brain
    Autonomic nervous system - information to and from organs and glands
        Sympathetic division - prepares for action
        Parasympathetic division - allows for relaxation

Central nervous system - all cells and cell parts which lie within brain
    Spinal cord - cable of long nerve fibers
        Reflexes - automatic actions that require no effort on our part
        Ascending nerve cells
        Descending nerve cells
    Hindbrain, Midbrain, Forebrain
        Hindbrain - contains medulla, pons, cerebellum
        Midbrain - reticular formation
        Forebrain - hypothalamus, thalamus, and cerebrum
            Cerebrum is divided into a left and right hemisphere
                Frontal lobe receives sensory impulses
                Occipital lobe receives impulses from eyes
                Parietal lobe responds to touch, pain, and temp.
                Temporal lobe receives sound and smell
    Right Brain/Left Brain
    Reticular Activating System activates all regions for incoming sensory impulses
    Limbic system plays a role in goal-directed behavior
        Performs many functions
    Voluntary muscle control - controlled by motor system
        Basal ganglia control "background" muscle tone
    Language control centers
        Broca's aphasia - slow or labored speech
        Wernicke's aphasia

Endocrine system regulates body chemistry
    Structure is made up of organs called endocrine glands, which release hormones
    Functions
        Pituitary gland is a pea-sized gland with wide-range of effects
        Gonads regulate secondary sex characteristics (sex hormones)
            Females gonads are the ovaries
            Males gonads are the testes
        Thyroid glands - produces thyroxin which transfers food to energy
            Metabolic rate determines how hungry you feel
        Parathyroid glands influence your energy level
        Adrenal glands produce adrenalin and noradrenaline
            Flight-or-fight response
        Pancreas controls the level of sugar in the blood
            Secretes insulin and glucagon into blood stream

Genetics - study of how traits are inherited, or passed on, from parent to child
        Genes are the cells from which we started
        Biochemistry of genes
                Genes are composed of DNA
                        Deoxyribonucleic acid (DNA) is the blueprint for life
                        Ribonucleic acid (RNA)
        Structure and function of chromosomes
                Sex cells (gametes) for female (ovum) and Male (sperm)
        Genetic laws of inheritance
                Single-gene traits
                Polygenic - more than one gene pair
                        Tyson's rat study - intelligence and genes
                Mutations - abnormal chromosome structures
                        Down syndrome - extra 21st chromosome

## DISCUSSION QUESTIONS AND EXERCISES

**Note: These questions and exercises ARE the learning objectives for this chapter. Answer them accurately in your own words and you will have mastered the most important material. We guarantee it.**

1. <u>History</u> <u>and</u> <u>Scope</u> of <u>Biopsychology</u>

a. What are **biopsychologists** concerned with?

b. Where and from who do the historical roots of **biopsychology** come?

c. What is **phrenology** and how is this theory received today?

30

d.  Describe some of the efforts that are being made to provide higher education to people with brain injuries.

2.  Neurons: <u>Basic</u> <u>Units</u> of <u>the</u> <u>Nervous</u> <u>System</u>

a.  Define the following basic components of the brain:

(1)  neuron cells

(2)  glial cells

b.  In what ways are neurons similar to all other cells and how are they not like other cells?

c.  Trace the path of an impulse as it travels through our neural network.

d.    How does **synaptic transmission** differ from axonal transmission?

e.    Identify the five parts of a synapse.

f.    Approximately how many chemical transmitters have been identified and what is the purpose of these neurotransmitters?

g.    Characterize each of the following chemical transmitters:

(1)    Acetylcholine

(2)    Catecholamines

(3)     Endorphins

3.    <u>Peripheral</u> <u>Nervous</u> <u>System</u>

a.    Identify the two subcomponents of the peripheral nervous system and give a brief description of each and its purpose.

4.    <u>Central</u> <u>Nervous</u> <u>System</u>

a.    What is the function of the **spinal cord**?  What relation does it have to **reflexes**?

b.    How do **ascending nerve cells** and **descending nerve cells** differ?

c. Identify the location, the components, and the function of the following:

   (1) Hindbrain

   (2) Midbrain

   (3) Forebrain

d. Match the following lobes with their appropriate function:

   (1) Frontal lobe          _____ sound and smell impulses
   (2) Occipital lobe        _____ touch, pain, and temperature
   (3) Temporal lobe         _____ receives impulses from the eye
   (4) Parietal lobe         _____ receives sensory impulses

e. What contributions to the field did Wilder Penfield make?

f.     How did Sperry's experiments further our understanding of the right and left brain division?  Why was this research considered so important?

g.     Characterize the functions of the **reticular activating system** and the **limbic system**?

h.     Identify some of the various methods to study the limbic system.

i.     The following three areas control voluntary muscle control.  Give a description of each and their function.

       (1)    Basal ganglia

(2) Motor cortex

(3) Cerebellum

j. Where is the language control center? Why do you suppose that it is located here?

5. <u>Endocrine</u> <u>System</u>

a. How is the **endocrine system** structured? Give a description of the path that the hormones take in the bloodstream.

b. Match the following glands with their appropriate function in the endocrine system:

(1) Pituitary gland        _____ the master gland
(2) Adrenal gland        _____ rate of energy transmission
(3) Gonads        _____ controls levels of sugar
(4) Pancreas        _____ produces sex hormones
(5) Parathyroid gland        _____ response to stress
(6) Thyroid gland        _____ general energy level

6. <u>Genetics</u>

a. What does **genetics** study?  What are and where are **genes**?

b. What are the genes composed of and give a brief definitions of these?

c. Identify the structure and function of chromosomes.

d. What did Mendel discover about the law of inheritance?

e.   Why are most human traits **polygenic**?  What relation to genes have with intelligence?

f.   What are mutations?  Give an original example not found in your book.

g.   What causes the abnormality **Down syndrome**?  How can it be detected early in pregnancy?

h.   Identify the three causes of mutations given to you in your textbook.

**POSTTEST**

1. Gall's theory of phrenology was important because it focused attention on the idea that:
   a. brain areas have specific functions
   b. genes can be transmitted from parent to child
   c. the head's shape is attractive to other members of the species
   d. the nervous system and the endocrine system interact

2. The concept that our body limits our experiences of the physical environment was first proposed by:
   a. Aquinas
   b. Aristotle
   c. Descartes
   d. Socrates and Plato

3. A single long fiber called the _____ carries messages away from the cell body.
   a. axon
   b. dendrite
   c. neuron
   d. synapse

4. The most common neurotransmitter is:
   a. acetylcholine
   b. catecholamine
   c. dopamine
   d. serotonin

5. If you cut your finger, your pain may be lessened by a shot of morphine. The transmitters that do this for you are called:
   a. catecholamines
   b. dendrites
   c. endorphins
   d. synapses

6. Body temperature is controlled by the:
   a. brain stem
   b. hindbrain
   c. hypothalamus
   d. thalamus

7. Scientists who study how the brain's properties relate to its physiological functions are called:
   a. biopsychologists
   b. cognitivists
   c. geneticists
   d. neuroanatomists

8. The specific pattern of genes inherited at conception defines an individual's:
   a. genotype
   b. ontogenic inheritance
   c. phenotype
   d. polygenic inheritance

39

9.      Which of the following statements describes the thalamus?
        a.      the brain's relay station
        b.      the connection between the cerebral hemisphers
        c.      the newest part of the brain, in evolutionary terms
        d.      the part of the brain that controls the pituitary gland

10.     The left and right cerebral hemisphere of the cerebrum are connected by the:
        a.      corpus callosum
        b.      longitudinal fissure
        c.      motor cortex
        d.      somatosensory cortex

11.     If someone taps your knee with a hammer, you kick.  This action required no
        conscious effort on your part and is known as a(n):
        a.      cognitive response
        b.      habit
        c.      nervous tic
        d.      reflex

12.     An important function of the reticular activating system is:
        a.      color perception
        b.      inducing sleep
        c.      lowering anxiety
        d.      selective attention

13.     Endocrine glands secrete:
        a.      enzymes
        b.      hormones
        c.      proteins
        d.      steroids

14.     Messenger molecules of _____ control how protein-contributing material
        is fashioned into protein chains.

15.     If the pituitary gland is the "master" of the _____ system, then the
        _____ could be characterized as the "master" of the central nervous
        system.
        a.      autonomic nervous; hindbrain
        b.      endocrine; forebrain
        c.      endocrine; midbrain
        d.      peripheral nervous; corpus callosum

16.     The basic units of heredity are called:
        a.      cells
        b.      chromosomes
        c.      genes
        d.      neurons

17.     The sex glands are called _____.

18.　A human gene is called _____ if it is located on the X chromosome in pair 23.
　　a.　abnormal
　　b.　mutated
　　c.　recessive
　　d.　sex-linked

19.　An example of a sex-linked trait is:
　　a.　color blindness
　　b.　diabetes
　　c.　Down Syndrome
　　d.　night blindness

20.　Dr. Owens has done an amniocentesis on a patient. The test shows that the fetus has 47 chromosomes. Dr. Owens would make which of the following conclusions?
　　a.　the baby is a girl
　　b.　the baby will be normal
　　c.　the baby will be overly aggressive
　　d.　the baby will likely suffer from Down's syndrome

21.　{T　F} When traits are determined by the action of more than one gene pair the trait is said to be multi-genetic.

22.　_____ allow neurons to communicate with other cells.
　　a.　Axons
　　b.　Dendrites
　　c.　Myelins
　　d.　Nodes

23.　A fatty covering around some axons is called:
　　a.　axon sheath
　　b.　the cell membrane
　　c.　the myelin sheath
　　d.　the nucleus

24.　The "on" state, or neural impulse, is called:
　　a.　depolarization
　　b.　membrane potential
　　c.　neuron potential
　　d.　polarization

25.　The junction between two neurons is called the:
　　a.　axon terminal
　　b.　dendrite
　　c.　synapse
　　d.　vesicle

26.　Jim just barely avoided a head-on collision on a narrow road. With heart pounding, hands shaking, and body perspiring, Jim recognizes that these are signs of the body's fight-or-flight response, which is controlled by the:
　　a.　empathetic division of the peripheral nervous system
　　b.　parasympathetic division of the autonomic system
　　c.　somatic division of the peripheral nervous system
　　d.　sympathetic division of the autonomic nervous system

41

27. A marathon runner may well experience a phenomenon known as "runner's high" because the pain of a long run may trigger the release of _____ in the brain.
    a. endorphins
    b. morphine
    c. naloxine
    d. placebos

28. Synapses using _____ can either transmit slow-acting or arousal messages.
    a. acetylcholines
    b. catecholamines
    c. endorphins
    d. vesicles

29. Which of the following responds to touch, pain, and temperature?
    a. frontal lobe
    b. motor cortex
    c. occipital lobe
    d. parietal lobe

30. {T  F} Motor responses of the body are controlled by the reflexes.

31. Estrogen is to _____ as testosterone is to _____.
    a. gonads; testes
    b. ovaries; testes
    c. ovaries; gonads
    d. testes; ovaries

32. The central nervous system is made up of the _____ and the _____.

33. In most people, the left hemisphere is to _____, as the right hemisphere is to _____.
    a. analyzing information; synthesizing information
    b. reading; writing
    c. recognition of complex geometric forms; recognition of faces
    d. visual recognition of words; recognition of speech sounds

34. Nancy was an accomplished theatrical dancer until she fell and hit her head. Since the accident, she has trouble making smooth and coordinated movements. The accident most likely caused damage to her:
    a. cerebullum
    b. hypothalamus
    c. medulla
    d. pons

35. William has been overweight since childhood. He diets frequently and can lose
    weight but always seems to gain it back because he is unable to control his
    eating. Hank may have a problem with his:
    a.    limbic system
    b.    pancreas gland
    c.    pituitary gland
    d.    thyroid gland

36. Tami felt and remembered nothing from the time she entered the operating
    room until she awoke several hours later. Her anesthetic blocked her:
    a.    limbic system
    b.    peripheral nervous system
    c.    reticular activating system
    d.    spinal cord's functionality

37. Broca's area is to _____ as Wernicke's area is to _____.
    a.    articulation; understanding
    b.    audition; vision
    c.    understanding; articulation
    d.    vision; audition

38. People with the pattern XXY suffer from:
    a.    diabetes
    b.    Down Syndrome
    c.    high levels of aggression
    d.    Klinefelter's syndrome

39. {T  F} The sympathetic system plays a role in emotions, memories, and
         goal-directed behaviors.

40. The most direct effect on a person whose nervous system was not producing
    sufficient amounts of acetylcholine or catecholamines would probably be which
    of the following difficulties:
    a.    aphasia
    b.    memory difficulties
    c.    migraine headaches
    d.    pain perception

41. After a tumor was surgically removed from Christine's forebrain, she could no
    longer tell if a stimulus was painful or pleasent. Which structure of Christine's
    forebrain was likely damaged?
    a.    hypothalamus
    b.    somatosensory cortex
    c.    temporal lobe
    d.    thalamus

42. Paralysis of the legs would most likely involve damage to:
    a.    ascending neurons
    b.    descending neurons
    c.    sensory and motor neurons in the autonomic nervous system
    c.    sensory and motor neurons in the peripheral nervous system

43

43. Jan has to monitor her diet very carefully to insure that she avoids certain foods and eats enough of others. Jan also has to take insulin orally. Jan likely has a malfunctioning _____ gland.
   a. pancreas
   b. parathyroid
   c. pituitary
   d. thyroid

44. During the relative refractory period:
   a. any impulse equal to or greater than the absolute threshold causes the neuron to fire
   b. no firing occurs
   c. the on-off firing process of the cell occurs less intensely
   d. the cell fires only in response to a very strong impulse

45. Neural messages travel faster on:
   a. myelinated axons
   b. non-myelinated axons
   c. the cell membrane
   d. the nodes of Ranvier

46. Kim wishes her letter to arrive quickly so she sends it by air mail while June is sending a letter by a much slower method, horseback. In this example, if the letter represents a neural message then sending it "airmail" would represent
   a. an axon with a myelin sheath
   b. a cell with many dendrites
   c. the relative refractory period
   d. the absolute refractory period

47. Which of the following best represents the correct order of events as they occur during depolarization?
   a. the charge reaches a critical value; cell membrane opens to positive sodium ions; the charge in cell changes from negative to positive
   b. the cell membrane closes to sodium, the charge reaches a critical value; charge in cell changes from negative to positive
   c. charge in cell changes from negative to positive; the charge reaches a critical value; the cell membrane opens to positive sodium ions
   d. the cell membrane opens to positive sodium ions; the charge reaches a critical value; the membrane closes to sodium

48. As you read this question, your eyes are gathering information, which is sent to and interpreted by your brain, and then your brain sends its commands to your muscles that will mark an answer. What is the correct type and order of nervous systems that are affected in this process?
   a. central; somatic; central
   b. somatic; central; autonomic
   c. somatic; somatic; autonomic
   d. somatic; central; somatic

49. The difference between the sympathetic and parasympathetic divisions is that
   a.  the sympathetic increases arousal while the parasympathetic controls relaxation
   b.  the sympathetic division goes to different glands than the parasympathetic division
   c.  the somatic nervous system regulates the sympathetic division while the autonomic nervous system regulates the parasympathetic division
   d.  the somatic nervous system regulates the parasympathetic division while the autonomic nervous system regulates the sympathetic division

50. A picture of a comb is shown to the right visual field of a split-brain patient. This means that the patient will point to the object with her _____ hand and _____ name the object.
   a.  right; cannot
   b.  right; can
   c.  left; cannot
   d.  left; can

## POSTTEST ANSWERS

1. a
2. d
3. a
4. a
5. c
6. c
7. a
8. a
9. a
10. a
11. d
12. d
13. b
14. RNA
15. b
16. c
17. gonads
18. d
19. a
20. d
21. F
22. b
23. c
24. a
25. c
26. d
27. a
28. b
29. d
30. F
31. b
32. brain; spinal cord
33. a
34. a
35. d
36. c
37. a
38. d
39. F
40. b
41. a
42. b
43. a
44. d
45. a
46. a
47. d
48. d
49. a
50. d

**PSYCH JOURNAL**

    **Please use the following pages to record your thoughts and feelings about the following questions.**

1.    Think about Agnes de Mllle's struggle to overcome brain damage as a mirror of her struggle to excel in dance and choreography. Write about it as a mirror of your struggle to excel and overcome physical disorders. This chapter should help you address specific questions including the following: Can you overcome physical disorders, such as headaches, with psychological changes, such as relaxation? Do your efforts to excel draw more on your left or right brain? Are your efforts limited or enhanced by being male or female?

_____

_____

_____

_____

_____

_____

_____

_____

_____

_____

_____

_____

_____

_____

_____

_____

_____

_____

_____

2. Do you know anyone who has ever suffered a stroke? How did the stroke affect him or her? How was s/he treated?

_____

_____

_____

_____

_____

_____

_____

_____

_____

_____

_____

_____

_____

_____

_____

_____

_____

_____

_____

_____

3.  After learning about the chemical reactions within the brain and nervous system, how do you view the use of "recreational" drugs and alcohol?

_____

_____

_____

_____

_____

_____

_____

_____

_____

_____

_____

_____

_____

_____

_____

_____

_____

_____

_____

4. Do you think there is a link between genetics and hemispheric specialization? Do you, your parents, and your siblings share many of the same abilities, hobbies, or occupations? How would you explain these similarities or differences?

_____

_____

_____

_____

_____

_____

_____

_____

_____

_____

_____

_____

_____

_____

_____

_____

_____

# Chapter 3 -- Sensation and Perception

## CHAPTER SUMMARY

1.  From a processing point of view, sensations correspond roughly to gathering information and perception corresponds to interpreting this information. From an experiential point of view, sensation corresponds to detecting something without knowing what it is. Perception corresponds to recognizing a specific object.

2.  Absolute threshold for the senses is the least amount of a stimulus that is detected 50 percent of the time. Signal-detection theory recognizes that sensitivity may be biased by decision criteria. Weber's Law states that a just-noticeable change in a stimulus magnitude is proportional to the original stimulus magnitude. Sensory adaptation is the decrease in the apparent intensity of a stimulus that is presented continuously. The first bite of a candy bar is always the most intense.

3.  In the eyes, two kinds of receptors in the retina, rods and cones, provide chemical reactions that change light into neural impulses. The impulses combine to send signals along pathways to the brain. The eyes can adjust their sensitivity to light. They become more sensitive in dim light through a process of dark adaptation. They become less sensitive in bright light through a process of light adaptation. Receptors in the retina are specialized so that they are more sensitive to some colors than others. According to the opponent-process theory, cells are joined in opposing pairs that interpret the colors we see. Specialized cells in the eye act as edge enhancers, feature detectors, and spatial frequency analyzers. Research on these cells has provided a foundation for understanding the neural foundation of pattern perception.

4.  We have perceptual constancies; that is, perceptions remain constant or stable when visual information changes. For instance, apparent size remains constant when we move further from an object even though movement shrinks the retinal image of the object. Unconscious inference theory and ecological theory offer two opposing explanations for perceptual constancies. We actively impose organization on what we see according to various rules. The Gestalt laws of organization include the law of proximity and the law of similarity. Others have proposed two new laws: the law of enclosure and the law of connectedness. The Gestalt school has also noted that we divide what we see into figure (an object) and ground (the background).

5.  We get a different view of the world from each eye. This difference is called binocular disparity. Monocular cues operate even when only one eye is used. The most important monocular cues include clearness, linear perspective and texture gradient. Two types of depth information are obtained through motion: motion parallax (the apparent pattern of object motion that is seen when an observer travels past those objects); and, kinetic depth effect (the apparent depth that depends on object motion).

6.  Illusions are perceptual distortions. In geometric illusions (such as the Muller-Lyer, Ponzo, Zollmer, and Titchener illusions) our eyes tell us one thing, but a ruler tells us something else. The autokinetic effect is the tendency for a stationary light viewed against darkness to look as if it is moving. This is particularly dangerous for pilots who have crashed while trying to fly next to what appeared to be another plane but that turned out to be a streetlight or marker buoy.

7.      People in cultures that are dominated by curved rather than straight lines are not very susceptible to the Muller-Lyer illusion.  The ability to see three dimensions in two-dimensional art may be influenced by the amount of exposure to this kind of art.  Pygmies who had never seen animals from a great distance thought faraway buffaloes were actually the size of ants.

8.      The ears convert sound waves into neural responses that we experience as hearing.  The external ear collects and focuses sound, causing the eardrum to vibrate.  Three delicate bones of the middle ear carry these vibrations to the oval window, which vibrates receptors in the cochlea, triggering nerve impulses to the brain.  Pitch (high notes versus low notes) is determined by the frequency of sound waves.  Loudness is determined by the amplitude of sound waves.  Tone quality is determined by how sound waves combine.  We hear the location of sound in part because we are sensitive to tiny sound differences between the two ears.

9.      The tongue has taste buds that convert chemical stimulation into neural responses, which we experience as taste.  Different regions on the tongue are sensitive to the four primary tastes: sweet, sour, salty, and bitter.  The passageway between the nose and the throat contains the odor-sensitive cells called olfactory cells.  Gases we breathe are dissolved in a fluid covering the receptors, causing the cells to send messages to the brain.  Olfactory cells also influence taste.  Skin receptors respond to pressure, temperature, and pain.  The receptors are viewed as information processing units in a system that includes pathways to the brain.  Body orientation is experienced through two kinds of senses.  Equilibrium, based on the body's reaction to gravity, is determined by the movement of fluid and tiny stones in the vestibular system and otolith structures.  Proprioception, the sense of where our body parts are, is determined by sense organs in joints, muscles, and tendons.


## KEY TERMS AND CONCEPTS

History and Scope of the Study of Sensation and Perception
        Processing and Experiential viewpoints
        Structuralists vs. Gestalt psychologists

Exploring the Limits of Our Sensations
        Psychophysics studies the relation between physical energy and psychological
                experience
                Absolute threshold is the least amount of stimulus energy that is detected
                Difference threshold is the smallest difference in intensity
        The Method of Constant Stimuli
                Used sometimes to overcome problems of other methods
        Method of Signal Detection
                Recognizes that sensitivity could be biased by decision criteria
                Signal-detection theory states that the probability of a response is
                        determined by both the senses and the process of decision making

Methods for Measuring Psychophysical Functions
        Special methods are used to constrain how people respond questions on relations
                Magnitude estimation
                Category judgment
                Cross-modality matching

52

Weber's Law states that a just-noticeable change in a stimulus is proportional to the original stimulus magnitude

Just-noticeable difference (JND) is the goal of the Weber fraction

Sensory Adaptation is the decrease in apparent intensity of a stimulus

Reduces sensitivity to continual information

## Seeing

Basic structures of the eyes

Cornea is the window of the eye

Aqueous humor carries nourishment for the eye

Glaucoma is caused by a breakdown in the recycling program

Vitreous humor is the gel that fills the eyes main chamber

Iris - the muscles behind the aqueous humor

Pupil

Lens - behind the pupil and focuses light

Near point accommodation - nearest point at which print is read

Sclera determines the eyes shape

Nearsightedness results when the eye is elongated

Farsightedness results when the eye is flattened

Retina contains receptors that respond to light

Blind spot - filled with optic nerves

Fovea is densely packed with nerves

Light Receptors: Rods and Cones

Located in the fovea

Use a chemical reaction (photopigments) to change light into impulses

Pathways to the Brain - the road the impulses travel

Light and Dark Adaptation

Light adaptation - decreased apparent intensity of light

Dark adaptation - increased apparent intensity of light

Both are examples of sensory adaptation

Color Vision and Color Blindness

Edges, Features, and Spatial Frequencies: the eye is not a camera

Perceptual Constancies

Lightness constancy is tendency for apparent lightness to remain constant

Lateral inhibition is a partial explanation for this

Size constancies

Law of size constancy

Objects size is constant regardless of changes in distance

Opposing theories

Unconscious inference theory

Ecological theory

Body size perception and eating

Used as a predictor of eating disorders

Shape constancies

Law of shape constancy

Objects shape is constant when the angle changes

Perceptual Organization

Gestalt laws of perceptual organization

Law of nearness or proximity

Law of similarity

Rock and Palmer's law of enclosure and law of connectedness

Our basic tendency is to divide what we see into a figure and ground

Figure-ground reversal (multistable perception)

Depth Perception
    Binocular disparity is the different view from each eye
    Seeing depth with one eye
        Monocular cues - used to see depth and objects distances
            Clearness
            Linear perspective (texture)
    Seeing depth through motion
        Motion parallax
        Kinetic depth effect - information about depth when objects move
Visual Illusions - perceptual distortions
    Geometric illusions
    Autokinetic effect
Perception and Culture

Hearing
    Basic Structures of the Ears
        External Ear
            Pinna is the elastic flap
            Ear canal is a tube-like passage
            Eardrum
        Middle Ear
            Contains hammer, anvil, and the stirrup
        Inner Ear
            Oval window
            Cochlea
            Basilar membrane
    Sound Transmission - bone conduction and air conduction
    Characteristics of Sound
        Frequency and Pitch
            Humans hear sounds ranging from 20 to 20,000 Hz
            Place theory
            Frequency theory
        Amplitude and Loudness
        Waveform and Tone Quality
    Locating Sounds
        Doppler shift - occurs when sound waves bunch up and then spread out

Other Senses
    Taste and Food Preferences
        Primary sense organ for taste is the tongue
            Papillae - small elevations (taste buds)
            Four primary tastes: sweet, sour, salty, bitter
        Smell - odor sensitive cells are called olfactory cells
    Touch: pressure, temperature and pain
    Body Orientation
        Equilibrium is difference between standing and sitting
        Proprioception is our sense of the position and motion of body parts

## DISCUSSION QUESTIONS AND EXERCISES

**Note:  These questions and exercises ARE the learning objectives for this chapter.  Answer them accurately in your own words and you will have mastered the most important material.  We guarantee it.**

1.   <u>History</u> <u>and</u> <u>Scope</u> <u>of</u> <u>the</u> <u>Study</u> <u>of</u> <u>Sensation</u> <u>and</u> <u>Perception</u>

a.   How are **sensations** and **perceptions** defined and contrasted?  What two viewpoints are given?

b.   What role do the Structuralists and Gestalt psychologists play in the study of sensation and perception?

c.   Define **psychophysics**.  Explain the three categories that psychophysical questions can be divided.

d.     Characterize each of the following methods given to you in your textbook:

   (1)     Method of Constant Stimuli

   (2)     Method of Signal Detection

e.     What is **signal-detection theory** and what does it depend on?  Identify some of its advantages.  How is it being used today?

2.     <u>Methods</u> <u>for</u> <u>Measuring</u> <u>Psychophysical</u> <u>Functions</u>

a.     Identify the methods psychologists have used to constrain how people respond to questions about psychophysical relationships.  Give an example of each.

b.   What is Weber's Law?  Why would someone want to use this principle or law?

c.   Identify **sensory adaptation** and define why it is useful.

3.   <u>Seeing</u>

a.   Match the following parts of the eye with its appropriate function:

| | | | |
|---|---|---|---|
| 1. | retina | _____ | window of the eye |
| 2. | aqueous humor | _____ | network of muscles |
| 3. | lens | _____ | nourishment for cornea |
| 4. | sclera | _____ | contains receptors for light |
| 5. | vitreous humor | _____ | gel that fills eyes' main chamber |
| 6. | iris | _____ | determines the eyes' shape |
| 7. | cornea | _____ | focuses light into back of eye |

b.   How are **nearsightedness** and **farsightedness** caused with relation to the **sclera**?

c.     Contrast **cones** and **rods**.  What role do each of them play in sight?

d.     Give an original example of **light adaptation** and **dark adaptation**.

e.     What is the opponent-process theory?  In what ways can it be used?

f.     Describe the process of edge enhancement.  How does it start and end?

g.     Identify and define the three different feature detectors given in your textbook.

h.     Give an example of a **lightness constancy**.  What may be a partial explanation for lightness constancy?

i.     What is the **law of size constancy**?  Explain how it happens.

j.     How does the **unconscious inference theory** differ from the **ecological theory**?

k.     Identify what role body size perception might have on your eating habits.

l.     What do the unconscious inference theory and ecological theory have to say about the **law of shape constancy**?

4.     Perceptual Organization

a.     Briefly characterize each of the following **Gestalt laws of organization**:

   (1)     law of nearness or proximity

   (2)     law of similarity

b.      What two new laws were proposed by Rock and Palmer?  Be sure to define each
        one of them.

5.  Depth Perception

a.      Identify **binocular disparity**.  How would you see disparity?  Do the example
        in the book and describe what you saw.  What popular toy uses this well?

b.      Identify the most important **monocular cue** and give a brief description of how
        artist use this cue.

c.      Define **motion parallax**.  Give an original example of this.

d.      Do the experiment with **kinetic depth effect** and give your thoughts on it below.

6.      <u>Visual</u> <u>Illusions</u>

a.      Why is the **autokinetic effect** dangerous?

b.      What role does culture play in perception?

7.      <u>Hearing</u>

a.      For each subdivision of the ear, identify the function and parts located in each subdivision:

        (1)     external ear

(2)     middle ear


(3)     inner ear


b.     What are some of the causes of **tinnitus**?  What are some of the treatments?


c.     How is sound transmitted?  Why do our voices sound weird when we listen to them on tape?


d.     What is **pitch** and **frequency**?  What sounds do humans hear?

e.     Briefly explain the **place theory** and **frequency theory**.

f.     What effects can loud sounds have on hearing?

g.     How can we distinguish between two sounds?  Identity two types of tones.

h.     Give an original example of the **Doppler shift**.  What happens when a sound comes from straight ahead and from one side?

8.   <u>Other</u> <u>Senses</u>

a.   What is the primary sense organ for taste?  Where are our **taste buds** located?

b.   Identify the four primary tastes?  What factors affect taste?

c.   Identify the passageway between the nose and throat.  Describe the process of smelling.

d.    Identify the different receptors in the skin.  Give a brief description of each one.

e.    What role do **equilibrium** and **proprioception** play in body orientation?

**POSTTEST**

1.    The science of _____ studies the relationship between physical
      energies and psychological experience.
      a.    psychoanalysis
      b.    psychobiology
      c.    psychophysics
      d.    physical psychology

2.    The least amount of a certain stimulus energy that can be detected is:
      a.    the absolute threshold
      b.    the differential threshold
      c.    the physical threshold
      d.    the psychological threshold

3.      When we look at a coin on a flat surface about a foot in front of us, it looks elliptical in shape. Nevertheless, we do not conclude that it IS elliptical in shape. We know that in some circumstances round things will look elliptical. This is an example of:
   a.   closure
   b.   figure-ground perception
   c.   good continuation
   d.   shape constancy

4.      The gel that fills the eye's main chamber and keeps it from collapsing is called the:
   a.   aqueous humor
   b.   glaucoma
   c.   sclera
   d.   vitreous humor

5.      The fact that your criterion for "hearing" mysterious noises at night may change after a rash of burglaries in the neighborhood can be best explained by:
   a.   Fechner's law
   b.   sensory adaptation
   c.   signal-detection theory
   d.   Weber's law

6.      {T F} Assuming the water temperature of your hot shower does not actually change, your feeling that the water is getting colder is due to sensory adaptation.

7.      Light receptors in the eye are called:
   a.   cones
   b.   lenses
   c.   pupils
   d.   rods

8.      The phenomenon of size constancy implies that:
   a.   the farther away an object is, the more we underestimate its true size
   b.   the perception of size is not related to the perception of distance
   c.   two objects will be perceived as the same size whenever they produce the same size retinal image
   d.   two objects will be perceived as being the same size though they produce different size retinal images

9.      One of the first tasks of perceptual organization is to distinguish figure from ground. Three qualities of figure-ground organization help us to do this. Which of the following is NOT one of these qualities?
   a.   the figure seems more dense and recognizable than the ground
   b.   the figure seems to be in front of the ground
   c.   the ground is formless and seems to extend behind the figure
   d.   the ground seems very dense compared to the figure

10.     In order to maximize visual acuity at night, you should:
   a.   blink your eyes many times to speed up dark adaptation
   b.   close one of your eyes
   c.   look directly at the object you want to see
   d.   turn your head at a slight angle to the object

11. The tube-like passage that funnels sound is called the:
    a. basilar membrane
    b. ear canal
    c. eardrum
    d. pinna

12. If you hear noises when there is no external source of sound, the condition is called:
    a. a delusion
    b. an hallucination
    c. bone conduction
    d. tinnitus

13. How are the opponent-process and trichromatic theories of vision and the place and frequency theories of hearing alike?
    a. the members of each of the two pairs complement each other
    b. the members of each of the two pairs contradict each other
    c. they are all attempts to explain color vision
    d. they are all attempts to explain how we register auditory frequency

14. If a person is presented with a series of pairs of light bulbs of different brightness and is asked whether the members of each pair differ in brightness, which of the following is being measured?
    a. the physical intensity difference between the two lights
    b. the subject's absolute threshold for brightness
    c. the subject's difference threshold for brightness
    d. the subject's visual acuity

15. The unit for measuring loudness is called a/an:
    a. amplitude
    b. decibel
    c. sound meter
    d. wave form

16. _____ results when an eye is elongated.

17. Compared to the rods, the cones are located _____ on the retina; and, to work best, the cones require _____ levels of illumination.
    a. more centrally; higher
    b. more centrally; lower
    c. more peripherally; higher
    d. more peripherally; lower

18. The sense of overall body orientation is called:
    a. equilibrium
    b. gustation
    c. otolith structures
    d. proprioception

19. Colored afterimages are best explained by which of the following?
   a. dual-receptor theory
   b. opponent-process theory
   c. primary process
   d. trichromatic receptor theory

20. A sense of the location of our body parts is called:
   a. cognitive maps
   b. equilibrium
   c. proprioception
   d. vestibular balance

21. The _____ detect changes in the position of the head.
   a. otolith structures
   b. proprioceptors
   c. semicircular canals
   d. vestibular equalizers

22. To impress your best friend with your knowledge of psychology, you ask him to increase the volume of your TV by "a JND," or "just noticeable difference." What is it you are asking your best friend to do?
   a. increase the volume so that is seems about twice as loud as before
   b. increase the volume so that it is equal to the absolute threshold
   c. increase the volume to the point where you detect that it is louder
   d. increase the volume until you reach the difference threshold

23. Research has found that Weber's Law:
   a. applies only to audition
   b. applies only to vision
   c. holds up well in all sensory modalities
   d. is not correct

24. The most widespread disease affecting vision is:
   a. cataracts
   b. glaucoma
   c. nearsightedness
   d. pink eye

25. If we were to examine the eye of an animal and found no cone cells, we might hypothesize that the animal:
   a. cannot perceive color
   b. cannot see in dim light
   c. has cataracts
   d. is blind

26. When the eye is flattened in shape like a vertical egg, the condition is called:
   a. color blindness
   b. farsightedness
   c. nearsightedness
   d. optic deterioration

27. Which of the following is not true of the electromagnetic spectrum?
   a.    it forms the entire range of electromagnetic radiation
   b.    the human eye can "see" X-rays
   c.    wavelengths corresponding to visible colors cover 90% of the spectrum
   d.    x-rays are much longer in wavelength than light waves

28. Which of the following theories state that receptor cells are joined in opposing pairs?
   a.    afterimage theory
   b.    opponent-process theory
   c.    proponent-process theory
   d.    trichromatic receptor theory

29. {T  F} Temporary deafness is a partial hearing loss caused by spending long periods of time in noises above 85 decibels.

30. The theory of color vision that states that the human eye has three types of receptors sensitive to different wavelengths is:
   a.    the opponent process theory
   b.    the trichromatic theory
   c.    the tridetector theory
   d.    the tricomponent theory

31. Which of the following does NOT help to explain how different frequencies are heard?
   a.    the acoustic theory of audiology
   b.    the frequency theory
   c.    the place theory
   d.    the volley principle

32. How might a person with red-green color blindness correctly interpret information provided by traffic lights?
   a.    ignore the traffic light, and pay attention to what other drivers in the area are doing
   b.    pay attention to the location of the lights; "stop" when the light at the top is on, "go" when the one at the bottom is on
   c.    there is no way; people with red-green color blindness should not drive
   d.    wear special glasses provided for this purpose

33. The combined response of different nerve fibers corresponds to the frequency of sound waves.  This is known as:
   a.    frequency theory
   b.    pitch theory
   c.    place theory
   d.    the volley principle

34. Which of the following describes how we hear according to the place theory?
   a. Different sound frequencies affect the intensity with which the membrane separating the middle ear from the inner ear vibrates, producing different pitches.
   b. Different sound frequencies vibrate different portions of the basilar membrane, producing different pitches.
   c. Our perception of pitch corresponds to the rate of frequency at which the entire basilar membrane vibrates.
   d. We perceive differences in pitch according to the number of hair cells that vibrate at any one time.

35. Older people are more likely than younger people to perceive foods as bland because:
   a. in older people, neurons in the area of the cortex responsible for processing gustatory information cease responding to inputs from the gustatory receptors
   b. older people lose their ability to replace lost or damaged taste buds
   c. they have adapted to gustatory stimuli as a result of many years of exposure
   d. they have less saliva than younger people

36. The tendency of some people to lean toward "false" as an answer to all true-false exam items could be considered to be an example of a _____.
   a. change in adaptation level
   b. correct rejection
   c. response bias
   d. false alarm

37. Which of the following occurs as cars zoom by on a race track?
   a. a loudness increase
   b. a phase shift
   c. a pitch decrease
   d. the Doppler shift

38. Which of following sense receptors is MOST important to taste?
   a. hearing
   b. olfactory cells
   c. proprioception
   d. vision

39. Which of the following states that thermal receptors are actually sensory nerves that detect contraction and dilation of blood vessels?
   a. paradoxical cold
   b. paradoxical heat
   c. specific receptor theory
   d. vascular theory

40. When binocular disparity is very great, a person might:
   a. not be able to perceive depth
   b. not see the image
   c. see double images
   d. see no change in vision

41.  The proprioceptors are located in:
     a.  joints, muscles, and tendons
     b.  the hypothalamus
     c.  the inner ear
     d.  the limbic system

42.  If a subject in a perception study were listening to sounds and asked to assign a number that is proportional to each sound's intensity, the method being used to measure the psychophysical function is:
     a.  category judgment
     b.  cross-modality matching
     c.  magnitude estimation
     d.  Weber's Law

43.  When you walk from a dimly lit room, say a theater, into a very bright area, say a sunny afternoon, your eyes should undergo a process called _____ adaptation. This process that takes a relatively _____ period of time to complete.
     a.  dark; long
     b.  dark; short
     c.  light; long
     d.  light; short

44.  You are standing in the middle of an abandoned long stretch of road in the desert. Of the following, which is the most important cue to realizing the road is actually a long stretch in the distance?
     a.  atmospheric perspective
     b.  interposition
     c.  linear perspective
     d.  movement parallax

45.  The small bones in the middle ear that transmit sounds to the inner ear are the _____, anvil, and _____.

46.  Dr. Engbring is interested in students' perceptions of the volume of rock music. Specifically, he lets students listen to music at a certain volume and then slowly increases the volume. He is interested in finding out how far he needs to turn up his stereo in order to find the level at which people realize that the music is louder than it was before. Dr. Engbring's work involves the concept of
     a.  absolute threshold
     b.  sensory adaptation
     c.  relative magnitude
     d.  difference threshold

47.  If a subject in a perception study were asked to assign a sound's intensity to an analogous level of a light's intensity, the method being used to measure the psychophysical function is
     a.  category judgment
     b.  magnitude estimation
     c.  signal detection
     d.  cross-modality matching

48.    When the light reaches the retina, it is first processed by _____, then the information travels to _____, then to _____ that sends signals out of the retina to the brain.
       a.    bipolar cells; rods and cones; ganglion cells
       b.    rods and cones; bipolar cells; ganglion cells
       c.    ganglion cells; bipolar cells; rods and cones
       d.    rods and cones; ganglion cells; bipolar cells

49.    The theory of size constancy that states that size constancy occurs because of unconscious accurate inferences about object size when we have accurate information about retinal image size and object distance is known as:
       a.    unconscious inference theory
       b.    constancy-adaptation theory
       c.    retinal image theory
       d.    ecological theory

50.    According to gate-control theory, the reason why rubbing the area surrounding a cut may reduce the pain is that it stimulates the:
       a.    slow fibers and thus narrows the pain gates
       b.    fast fibers and thus narrows the pain gates
       c.    slow fibers and thus widens the pain gates
       d.    fast fibers and thus widens the pain gates

## POSTTEST ANSWERS

1.   c
2.   a
3.   c
4.   d
5.   c
6.   T
7.   a
8.   d
9.   d
10.  d
11.  b
12.  d
13.  a
14.  c
15.  b
16.  Nearsightedness
17.  a
18.  a
19.  b
20.  c
21.  c
22.  c
23.  c
24.  a
25.  a
26.  b
27.  b
28.  b
29.  F
30.  b
31.  a
32.  b
33.  d
34.  b
35.  b
36.  c
37.  d
38.  b
39.  d
40.  c
41.  a
42.  c
43.  d
44.  c
45.  hammer; stirrup
46.  d
47.  d
48.  b
49.  a
50.  a

**PSYCH JOURNAL**

   **Please use the following pages to record your thoughts and feelings about the following questions.**

1.   Think about the many ways in which Diane Fossey's experiences illustrate principles of sensation and perception.  Write about how your everyday experiences illustrate principles of sensation and perception.

_____

_____

_____

_____

_____

_____

_____

_____

_____

_____

_____

_____

_____

_____

_____

_____

_____

2. Think about one of memories: a family trip; your senior prom; your favorite Christmas custom. Describe that memory in as much sensory detail as possible, including information about your physical, visual, auditory, olfactory, and taste *sensations* and *perceptions*.

_____

_____

_____

_____

_____

_____

_____

_____

_____

_____

_____

_____

_____

_____

_____

_____

_____

_____

_____

_____

3.   Most of us have seen a variety of optical illusions.  Using a specific example (one not in the book), discuss how one of these illusions works.

_____

_____

_____

_____

_____

_____

_____

_____

_____

_____

_____

_____

_____

_____

_____

_____

_____

_____

_____

_____

_____

4. Think about playing baseball or driving a car, especially at night. How would such automatic, seemingly simple everyday tasks be affected if you had poor depth perception?

_____

_____

_____

_____

_____

_____

_____

_____

_____

_____

_____

_____

_____

_____

_____

_____

_____

_____

_____

_____

# Chapter 4 -- Alternate States of Consciousness

**CHAPTER SUMMARY**

1.      John Lock defined consciousness as "the perception of what passes in a man's own mind."  Today's scientists refer to alternate states of consciousness as "specific patterns of physiological and subjective responses."

2.      Through self-experience individuals explore their own responses.  Questionnaires and interviews allow researchers to create questions and collect responses.  Laboratory experiments involve the use of subjective and objective measures.  The clinical cases method focuses on individuals with real problems.

3.      Sleep, a period of rest for the body and mind, has five stages that are distinguished by distinct patterns of brain waves and other psychological responses.  Stage 1 is marked by the appearance of theta waves.  Stage 2 is distinguished by the onset of sleep spindles.  Sleep spindles diminish in stage 3 and are replaced in stage 4 by delta waves.  Heart rate, respiration, and muscle tension steadily decline during the first four stages.  After completing stage 4, people go back through stages 3 and 2 before entering the REM stage.  Dreaming occurs during REM sleep.  Most adults sleep between 6 and 9 hours, 25 percent of which is spent in REM sleep.  Prolonged sleep deprivation leads to irritability, fatigue, poor concentration, memory failure, and reduced coordination.  Selective deprivation of REM sleep suggests that it is especially important in solidifying memories for skills learned the day before.

Dreams are a series of images, thoughts, and emotions occurring during sleep.  According to the perceptual release theory, brain activity during REM causes dreams.  Freud believed that dreams are vehicles through which individuals can express their unacceptable impulses in disguised or acceptable forms.  According to her cognitive problem solving theory, Rosalind Cartwright maintains that dreams can be used to solve everyday problems.  Our sleep-wake cycles are part of our circadian rhythms.  If work or play disrupts the natural rhythm of our sleep cycles, our performance suffers.  Rotating work shifts and insuring that individuals get at least 6 hours' sleep each night helps avoid the effects of sleepiness.  Experts disagree on how much sleep is necessary for each individual.  Sleep disorders such as insomnia and hyperinsomnia affect about 15% of the population.  Treatment for nightmare sufferers include desensitization and rehearsal.

4.      Daydreams are "thoughts that divert attention away from an immediately demanding task."  They are not characterized by unique physiological patterns.  The type of daydream you have may affect your behavior, particularly in relation to alcohol or drug abuse.

5.      Four common depressants are alcohol, sedatives, tranquilizers, and nicotine.  They all inhibit neural activity.  Alcohol reduces social inhibitions.  Drinking rate, stomach content, and gender affect the amount of alcohol that reaches the brain.  Sedatives are used to induce sleep and in lower doses, they induce a relaxed feeling of well being, poor concentration, and impaired motor coordination.  Tranquilizers such as Librium and Valium help relieve anxiety.  Nicotine is often classified as a stimulant because it initially stimulates the nervous system, but we classify it as a depressant

because it ultimately suppresses the central nervous system. Narcotics (also known as opiates) induce sleep and relieve pain. The pain relieving effects of opiates is achieved by boosting the effects of endorphins.

Stimulants include caffeine, amphetamines, cocaine, and "ecstasy". The benefits of caffeine are increases in both mental and physical work capacity. The costs are that higher and higher doses are required to achieve this lift. One who tries to stop taking caffeine experiences withdrawal symptoms. Amphetamines stimulate the central nervous system by enhancing the effects of excitatory neurotransmitters. They enhance the feeling of alertness and speed up reaction times, but they also increase errors in decision making. Cocaine blocks the removal of norepinephrine and dopamine; it arouses self-confidence and optimism, but also can lead to agitation, sleeplessness, paranoia, sudden depression and permanent brain damage. Ecstasy (MDMA) is a designer drug that can cause hallucinations and permanent brain damage. Hallucinogenic drugs include marijuana, mescaline, psilocybin, LDS, and PCP. Drug-induced hallucinations seem to have four stages. This is true across cultures.

6.    Hypnotists can control both physiological and subjective responses in susceptible subjects. Physiological measures do not distinguish hypnosis from waking consciousness, which leaves room for debate about whether hypnosis induces an alternate state of consciousness. While scientists debate this, they agree that hypnotists can influence judgment, suggestibility, relaxation, attention, and memory, both during and after hypnosis. Important applications include the reduction of pain.

7.    Meditation is a process combining mental and physical acts to achieve an alternate state of consciousness. This process can reduce oxygen consumption, respiratory rate, heart rate, blood pressure, and muscle tension. It also induces subjective responses characterized by feelings of calm and well-being. Meditation is associated with reduced drug consumption.

8.    Sensory deprivation is achieved through experiments where sensory stimulation is drastically reduced in quantity and intensity and in real settings such as isolation cells in prisons. Sensory deprivation changes physiological responses, but not in a way that distinguishes it as a unique state of consciousness.

9.    ESP is the reception of information by means other than our usual senses of hearing, sight, taste, touch, and smell. PK is direct mental influence over physical objects or processes. Objective evidence of such powers does not meet the usual scientific standard, but many scientists are continuing to seek better evidence.

10.    Near-death experiences occur when people have lost all measurable vital signs and then are revived. Some of them report having subjective responses while they are supposedly dead. Some elements of these responses are experiences that are impossible to describe, hearing their doctors pronounce them dead, seeing dead friends and relatives, and accurately reporting events that occurred while they were supposedly dead. Near-death experiences are very similar, and many aspects of these experiences turn up repeatedly.

**KEY TERMS AND CONCEPTS**

History and Scope of Studying Alternate States of Consciousness
 Study of **consciousness** has roots in common with study of the body
 Important milestones
  John Locke's definition of consciousness
  Wilhelm Wundt's consciousness of surroundings
 Today it is defined as **alternate states of consciousness**
  Unique pattern of physiological and subjective responses
Methods for Studying States of Consciousness
 Self-Experience Approach
  Researchers experience the states themselves
  Has been applied to consciousness-altering drugs
 Questionnaires and Interviews
  Control exact wording in questionnaires
  Learn more about a person in interviews
 Laboratory Experiments
  Artificial means of studying alternative states of consciousness
 Clinical Case Method
  Study of someone who has a real problem
  Able to probe actual cause and remedy the cause
  Generalizability is a concern for this method


Sleep and Dreams
 Sleep is the period of rest for the body and mind
  Bodily functions are partially suspended
  Sensitivity to external stimuli is diminished
 Dreams are a series of images, thoughts, and emotions during sleep
 Physiological patterns during sleep
  Five stages of sleep
   Characterized by changes in brain waves
   Stage 1 - theta waves
   Stage 2 - **sleep spindles** occur
   Stage 3 - diminishing of sleep spindles
   Stage 4 - delta waves
   People go back to 3 and 2 before entering **REM** sleep
  Sleep deprivation causes **REM rebound**
 Subjective patterns during sleep
  The voices or visions we see during sleep are hallucinations
   Stage 1 hallucinations are called hypnagogic images
    These images are normal
  Dreaming occurs during REM sleep
 Why We Sleep
  Sleep deprivation suggests that sleep repairs and restores us
   Prolonged deprivation leads to irritability, fatigue
  Selective deprivation of REM sleep
   REM sleep is important for memory for skills learned day before
   REM promotes new connections between neurons
   REM affects our mood
 Why We Dream
  Perceptual release theory states that dreams are caused by actions and
   reactions in the brain
   This theory explains the physiological answer

Psychological importance of dreams
    Freud used interview procedures to record dreams
    Two levels in dreams according to Freud
        **Manifest content** - remembered material in dreams
        **Latent content** - hidden meaning of the dream
    Cartwright's **cognitive problem-solving theory**
        Dreams are important thought processes
The Rhythmic Nature of Sleep
    Circadian rhythm is process that repeats about every 24 hours
        Affected by cycles of light and dark
Sleepiness, and Performance at Work and School
Sleep Disorders
    Insomnia is abnormal wakefulness
    Hypersomnia is abnormal sleepiness
    15% of Americans fall in to at least one or both categories
    6 to 9 hours of sleep are usually all that is needed
        Apnea is the inability to breath properly during sleep
            Occurs in middle-aged adults and infants
        Narcolepsy is a serious form of hypersomnia
    Biofeedback and desensitization are used to treat these disorders

Daydreams
    Thoughts that divert attention away from an immediately demanding task
    We all daydream
    Physiological patterns
        No unique physiological patterns to daydreaming
        Blank (or unfocused) stare of the eyes is characteristic
        Females report it more than males
    Subjective patterns
        Three categories of daydreams
            Unhappy - fantasies involving guilt, fear of failure
                These people are more likely to drink alcohol
            Uncontrollable - fleeting, anxiety-ladden fantasies
            Happy - fantasies for planning future activities
                May help prevent alcohol or drug abuse

Drug-Induced States
    **Psychoactive drugs** are drugs that produce subjective effects
        Four categories
            Depressants - slow body functions or calm
                Common ones are alcohol, sedatives, nicotine
            Narcotics (opiates) - relieve pain and induce sleepy relaxation
                Include opium, morphine, heroin, and codeine
            Stimulants - excite body functions
                Common one is caffeine, also uppers, downers, and speed
            Hallucinogens - cause hallucinations
                Marijuana, LSD are common in this category
    Caffeine is the most widely used psychoactive drug
        Used to increase energy, reducing inhibitions, and relaxing
    Culture and personality influences our subjective responses to drugs
    Neurotransmitters play a major role in the effects of drugs

Hypnosis
> A state of consciousness induced by the words and actions of a hypnotist
> Used with pain control, heart attack patients, and those with allergies
> Hypnosis is a controversial issue about how it works
> Physiological patterns
>> There are no consistent physiological patterns
> Subjective patterns
>> Influences judgment and suggestibility
>>> **Posthypnotic suggestion** occur after sessions
>>> **Hypnotic susceptibility tests** predict willingness for hypnosis
>>>> Useful in medical treatments
>> Hypnotism influences all of the following:
>>> Relaxation - increase or decrease of tension
>>> Attention - division of attention
>>> Memory - increase or decrease in ability to recall facts
>>>> **Posthypnotic amnesia** - subject can't recall certain facts
>>>> **Hypermnesia** - increased recall resulting from hypnosis

Meditation
> This involves a series of mental and physical acts aimed at achieving an alternative state of consciousness
> Zen and yoga are the two Eastern forms of meditation practiced often
> Physiological patterns
>> **Relaxation response** refers to the physiological patterns observed
>>> Decreases oxygen consumption, respiratory rate
>>> Increases alpha waves
>>> Opposite of **fight-or-flight response**
> Subjective patterns
>> No encounters with subjective patterns are usually seen with meditation

Sensory Deprivation
> Sensory stimulation is drastically reduced in quantity and intensity
> For short periods it could induce relaxation and calm
>> Used for drug addiction, hypertension, and alcoholism
> Physiological patterns
>> Breath rate, heart rate, blood pressure changes are observed
> Subjective patterns

Hidden Mental Powers
> **Extrasensory perception (ESP)** is the reception by means other than usual ones
>> Three kinds of ESP
>>> Precognition - perception of future thoughts
>>> Clairvoyance - perception of objects without sensory stimulation
>>> Telepathy - perception of another's mental state or emotion
> **Psychokinesis (PK)** is direct mental influence over physical objects
> Parapsychologists study both of these

Near-Death Experiences
> Physiological patterns
>> There are no patterns here because the subject is dead
>> There is thought to be some undetectable source of biological activity
> Subjective patterns
>> 15 common elements found in near-death experiences

## DISCUSSION QUESTIONS AND EXERCISES

**Note: These questions and exercises ARE the learning objectives for this chapter. Answer them accurately in your own words and you will have mastered the most important material. We guarantee it.**

1. <u>History</u> <u>and</u> <u>Scope</u> of <u>Studying</u> <u>Alternate</u> <u>States</u> of <u>Consciousness</u>

a. Give a brief description of the history of consciousness. Identify some of the important milestones in these studies.

b. Define **alternate states of consciousness**. How do scientists identify each state? Give some examples.

2. <u>Methods</u> <u>for</u> <u>Studying</u> <u>States</u> of <u>Consciousness</u>

a. What is the self-experience approach to studying subjective patterns in various states of consciousness? What are some of the dangers of using this approach?

b.      How are questionnaires, interviews and laboratory experiments used in studying various states of consciousness?  Give an original example of each method.

c.      What are some of the limitations of the clinical case method?  How can they be avoided?

3.      <u>Sleep</u> <u>and</u> <u>Dreams</u>

a.      Define **sleep** and **dreams**.

b.      How are sleep patterns and dreams measured in the laboratory?

c.  Characterize each of the five stages of sleep, giving brief definitions of each one and the period of time each one occurs in sleep.

d.  What effect does sleep deprivation have on the pattern of stages that occurs?

e.  Are **hypnagogic** images normal? Why are people so embarrassed when they have them?

f.  What characteristics of dreaming have been recorded in research with dreams?

g.      Identify the various consequences of sleep deprivation.  What sorts of behavior occur?  What effect does deprivation of REM sleep have on us?

h.      What is the relationship between learning and REM sleep?  What effect does REM sleep have on our moods?

i.      According to the **perceptual release theory**, what is the connection between REM sleep and dreaming?

j.      Give a brief definition of Freud's **manifest content** and **latent content**. Provide an original example of a dream that would include both of these and identify them in your example.

k.     Cartwright has proposed the **cognitive problem-solving theory**.  How does this theory explain how we solve everyday problems?  Write down the methods given to help you remember your dreams.

l.     What are **circadian rhythms** and when do they occur?  Biologically, how is it regulated?

m.     How does sleepiness affect us at work and at school?  What could be done to reduce sleepiness?

n.     Identify and describe the two broad categories of sleep disorders.  How many people are estimated to suffer from these sleep disorders?

o.    What is a serious form of both insomnia and hypersomnia?  What are some of the methods used to control both of these?

4.    <u>Daydreams</u>

a.    What are **daydreams**?  Are there any physiological patterns to daydreams? Why or why not?

b.    Characterize each of the following three categories of daydreams:

(1)    unhappy

(2)    uncontrollable

(3)    happy

c.    What is the relationship between alcohol abuse and daydreams?

## 5. Drug-Induced States

a. Identify the four categories of **psychoactive** drugs. What is the most widely psychoactive drug used in America?

b. How does your culture or your personality influence our subjective responses to psychoactive drugs?

c. Identify four common depressants. What is their main psychoactive effect? Biologically, how do depressants affect us?

d.   What are some common narcotics and stimulants?  Identify the most common stimulant.  Why do people use these types of psychoactive drugs?

e.   Give a few of the common hallucinogenic drugs provided in the textbook and a brief description of each of the four stages of a drug-induced hallucination.

6.   Hypnosis

a.   What is the controversy over hypnosis?  What has hypnosis been used for in the past?  Have you ever been hypnotized?

b.   Give a brief description for each of the following changes that a hypnotists can cause:

   (1)   Hypnosis influences judgment and suggestibility

(2)     Hypnotism influences relaxation

(3)     Hypnotism influences attention

(4)     Hypnotism influences memory

7.     <u>Meditation</u>

a.     Define **meditation** and give two common forms of meditation practiced most often.  Briefly describe **transcendental meditation**.

b.     What are some of the physiological patterns seen during meditation?  Are there any subjective patterns?  What is the goal of meditation?

8. <u>Sensory</u> <u>Deprivation</u>

a. Define **sensory deprivation**. Give an example how this might be done and what result you might expect from such a study.

b. What are some typical physiological and subjective responses during sensory deprivation?

9. <u>Hidden</u> <u>Mental</u> <u>Powers</u>

a. What is **extrasensory perception (ESP)**? Give an example. How is it typically tested? Do you think that it "really" exists? Support your answer to the latter part of this question.

b.  What is **psychokinesis**?  Give an example.

10.  <u>Near-Death</u> <u>Experiences</u>

a.  When someone has a near-death experience, what are the physiological patterns? How does Moody account for near-death experiences?

b.  Summarize some of the fifteen common elements found in near-death experiences.

**POSTTEST**

1.  The perception of what passes in a person's own mind is called _____ according to Locke.
    a.   consciousness
    b.   dreaming
    c.   fantasy
    d.   unconsciousness

2.  Differing mental states as measured by specific patterns of physiological and subjective responses are called:
    a.   abnormal states of consciousness
    b.   alternate states of consciousness
    c.   fantasy
    d.   hallucinogenic states

3.  Which of the following is NOT listed as a method of studying states of consciousness:
    a.   correlational studies
    b.   laboratory experiment
    c.   questions and interviews
    d.   self-experience

4.  A period of rest for the body and mind during which bodily functions are partially suspended is called:
    a.   dreaming
    b.   fantasy
    c.   meditation
    d.   sleep

5.  According to Freud, the dream that is remembered is the:
    a.   latent content
    b.   manifest content
    c.   opposite of what the dream means
    d.   symbolic content

6.  The series of images, thoughts, and emotions that occur during sleep are called
    _____.

7.  Which of the following is likely to be a disadvantage when the laboratory experiment method is used to study alternate states of consciousness?
    a.   factors that might influence the states can be controlled
    b.   objective as well as subjective data can be collected
    c.   subjects can be selected for both control and generalization purposes
    d.   the alternate state being studied may be affected or tainted by the setting or the manipulations used

8.  {T  F} Stage 2 of sleep is marked by the onset of REM sleep.

9. Without the aid of exposure to the true day-night cycle (e.g., working and living underground) we would adopt a circadian rhythm that is
   a. a little longer than 24 hours
   b. a little shorter than 24 hours
   c. 48 hours long
   d. irregular, leading to possible derangement

10. The sharp increase of REM sleep after going through a period of sleep deprivation is called:
    a. alpha rhythms
    b. delta waves
    c. REM rebound
    d. REM spindles

11. Eric has sleep apnea. This means
    a. he awakens several hundred times each night because of difficulty breathing
    b. he falls asleep uncontrollably
    c. he has difficulty falling asleep
    d. he wakes up several hours early and then has difficulty going back to sleep

12. If you have high blood pressure, you might want to try which of the following in place of, or in conjunction with, medication?
    a. biofeedback
    b. biorhythms
    c. perceptual release
    d. psychoanalysis

13. Marcia suddenly jolted awake. Her heart was pounding, she was sweating, and she couldn't remember what woke her. Marcia probably:
    a. experienced sleep tremors
    b. had a hypnagogic image
    c. was in REM
    d. was in stage 1 sleep

14. Joe often falls asleep standing up. He may be suffering from:
    a. apnea
    b. narcolepsy
    c. nightmares
    d. sleep spindles

15. The brain activity observed during REM sleep:
    a. can indicate the content of dreams
    b. is extremely slow and regular
    c. is extremely fast and irregular
    d. is similar to that observed in wide-awake subjects

16. Hallucinations that occur during the drowsy interval before sleep are called:
    a. hallucinations
    b. hypnagogic images
    c. hypnotic images
    d. hypnotic trances

17.     {T  F} Drugs that produce subjective effects are called psychedelic drugs.

18.     According to the text, which of the following has been supported by research?
        a.      dreams can help anyone to solve a problem
        b.      dreams can be influenced by external stimulation
        c.      dreams reflect unconscious, unacceptable impulses
        d.      the harder you work during the day, the more you dream at night

19.     Sleep spindles, which appear against a background of mixed, mostly lower frequency EEG activity, are characteristic of:
        a.      REM sleep
        b.      stage 1 sleep
        c.      stage 2 sleep
        d.      stage 4 sleep

20.     A state of consciousness induced by words and actions of someone whose suggestions are readily accepted by the subject is called:
        a.      a dream
        b.      hypnosis
        c.      mesmerism
        d.      narcolepsy

21.     Increased recall that can result from hypnosis is called:
        a.      hypermnesia
        b.      mnemonics
        c.      REM recall
        d.      sensory increase

22.     The relationship between sensory deprivation and physiological patterns is:
        a.      clear; deprivation produces a sequence of physiological changes
        b.      clear with respect to involuntary bodily processes such as heart rate, but unclear with respect to brain waves
        c.      indistinct; deprivation produces physiological changes but not a pattern of them
        d.      poor; deprivation produces no physiological changes

23.     Which of the following characteristics are shared by marijuana and LSD?
        a.      they both produce "flashback" experiences
        b.      they are both classified as hallucinogenic substances
        c.      they both combat boredom
        d.      very small amounts produce very intense reactions

24.     {T  F} To Freud, the latent content of a dream predicts future events.

25.     According to Hobson and McCarley, we can "see" and "feel" during our dreams because:
        a.      pyramidal cells activate parts of our brain associated with vision and emotion
        b.      the mind functions apart from the sleeping parts of the brain
        c.      we perceive an external reality that is hidden from us during waking hours
        d.      we perceive an internal reality that is hidden from us during waking hours

26.  If you interpreted Sam's dream of arguing with a larger-than-life faceless authority figure as an attempt on his part to decide which approach to take in convincing his father of the merits of his future plans, you would be using the:
a.  activation-synthesis theory of dreams
b.  cognitive problem-solving theory of dreams
c.  neural overflow hypothesis of dreams
d.  wish-fulfillment theory of dreams

27.  Dream research suggests that our dreams are:
a.  dangerous if the content is not controlled
b.  nightmares
c.  predictions of the future
d.  rather ordinary

28.  About 10 minutes after going to bed, Alice suddenly jerked because she felt like she was falling.  Alice's jerking movement was probably associated with:
a.  a delusion
b.  a hypnagogic image
c.  a dream
d.  REM sleep

29.  {T  F} During REM sleep, brain waves show the characteristic low frequency high amplitude "delta" wave pattern.

30.  A mantra is a(n):
a.  auditory stimulus that makes meditation easier to achieve
b.  secret phrase repeated during meditation
c.  visual focal point that aids the meditative process
d.  visual image occurring during stage 1 of meditation

31.  A form of hypersomnia in which there are recurrent attacks of an uncontrollable desire for sleep is called:
a.  amnesia
b.  apnea
c.  narcolepsy
d.  sonorism

32.  {T  F} The most common type of hallucinations involve people "hearing" voices that are not there.

33.  Stimulant is to depressant as:
a.  alcohol is to barbiturates
b.  caffeine is to amphetamines
c.  cocaine is to alcohol
d.  mescaline is to barbiturates

34.  Drugs that excite body functions are called _____.

35.  In which of the following situations is a person most likely to daydream?
a.  when attempting to do a task that one has not previously done
b.  when learning a new task
c.  when one is doing a very rewarding activity
d.  while doing a task that has been successfully accomplished before

36. If you have never experienced a hypnagogic image it may be because:
   a. the images are associated only with psychiatric symptoms
   b. you pass through stage 1 too quickly
   c. you pass through stage 2 too quickly
   d. you think those images are abnormal

37. {T F} You are sitting in a class thinking about the novel you are planning to write someday. You are having a hypnagogic image.

38. There is a positive correlation between alcohol abuse and _____.
   a. infrequent or no daydreaming
   b. unhappy daydreaming
   c. uncontrollable daydreams
   d. happy daydreams

39. Today consciousness and alternate states are known to possess both physiological and subjective response components. Who first proposed that people be carefully observed to clarify the relationship between the body and consciousness?
   a. Hippocrates
   b. Locke
   c. Plato
   d. Wundt

40. Mary took LSD in an experiment. She reported looking down on her grandmother's house from a tree where she used to play. What stage of drug-induced hallucinations was she in?
   a. 1st
   b. 2nd
   c. 3rd
   d. 4th

41. Scientists who study ESP and PK are called _____.

42. Bill is in a state of alpha rhythm, with decreased heart rate and oxygen consumption. Bill is probably:
   a. hypnotized
   b. in a drug-induced trance
   c. in REM sleep
   d. meditating

43. Which of the following is NOT listed as a subjective or behavioral pattern associated with hypnosis?
   a. hypnotism influences judgment and suggestibility
   b. hypnotism influences memory
   c. hypnotism influences relaxation
   d. hypnotism influences sexual activities

44. The most common variety of insomnia is:
    a. difficulty falling asleep
    b. waking up hundreds of times a night for a brief second or two because of difficulty breathing
    c. waking up several hours early and being unable to go back to sleep until it is time to get up
    d. waking up several times a night for no apparent reason and then being unable to go back to sleep quickly

45. Tom knows his dad is going to call him in three days, even though no one has heard from him in 10 years. If true, Tom may have which ability?
    a. clairvoyance
    b. precognition
    c. psychokinesis
    d. telepathy

46. Some scientists believe that the best way to understand the relationship between physiological response patterns and consciousness is to study the link between _____ and specific contents of our consciousness.
    a. specific neural activity
    b. behavior
    c. our thoughts
    d. specific emotional patterns

47. While he is sleeping Neil feels a great deal of anxiety, his breathing becomes heavy and labored, and he feels as though he cannot move. Neil is experiencing
    a. REM rebound
    b. hypnagogic images
    c. dream imagery
    d. night terrors

48. Landa has been suffering from frequent nightmares. The doctors at the sleep clinic advise her to think about a new ending to her dreams and then write them down. She was then told to visualize her revised dreams while she was in a relaxed state. What kind of treatment is Landa undergoing?
    a. biofeedback
    b. rehearsal
    c. free association
    d. desensitization

49. Jennifer is undergoing hypnosis to help her quit smoking. The hypnotist tells Jennifer that she will become ill every time she picks up a cigarette after the hypnosis has ended. This is an example of:
    a. self-hypnosis
    b. hypermnesia
    c. posthypnotic suggestion
    d. posthypnotic amnesia

50. One reason that many scientists are skeptical about ESP and PK is that
   a. these topics have never been scientifically studied.
   b. many ESP and PK experiments cannot be consistently replicated.
   c. research has already conclusively proven that these abilities do not exist.
   d. most ESP and PK studies have been done in tightly controlled experiments which have little relation to real world occurrences.

## POSTTEST ANSWERS

1.     a
2.     b
3.     a
4.     d
5.     b
6.     dreams
7.     d
8.     F
9.     a
10.    c
11.    a
12.    a
13.    a
14.    b
15.    d
16.    b
17.    F
18.    b
19.    c
20.    b
21.    a
22.    c
23.    b
24.    F
25.    a
26.    b
27.    d
28.    b
29.    F
30.    b
31.    c
32.    T
33.    c
34.    stimulants
35.    d
36.    b
37.    F
38.    b
39.    a
40.    c
41.    parapsychologists
42.    d
43.    d
44.    a
45.    b
46.    a
47.    d
48.    b
49.    b
50.    b

## PSYCH JOURNAL

**Please use the following pages to record your thoughts and feelings about the following questions.**

1.  Use George Ritchie's story and the chapter as a guide for writing about how alternate state experiences have affected your life. Ritchie believed that his near-death experience changed his life for the better. Similarly, this chapter reviews evidence that we all can benefit from alternate states induced by "natural heights" such as happy day-dreams, hypnagogic images, dreams, and meditation. Have alternate experiences affected your life for better or worse?

_____

_____

_____

_____

_____

_____

_____

_____

_____

_____

_____

_____

_____

_____

_____

_____

_____

_____

_____

2. Have you ever pulled an "all-nighter" when studying for exams? How did you do on the test? What is a more effective and sane way to maximize your retention of material?

_____

_____

_____

_____

_____

_____

_____

_____

_____

_____

_____

_____

_____

_____

_____

_____

_____

_____

_____

3. Do you smoke, drink coffee, or drink an occasional beer? Do you think that alcohol, tobacco, caffeine and other common substances should be controlled by law because they are psychoactive drugs? Why or why not? What differences do you see between these substances and "real" drugs like cocaine, marijuana, uppers, etc?

_____

_____

_____

_____

_____

_____

_____

_____

_____

_____

_____

_____

_____

_____

_____

_____

_____

_____

_____

_____

4. Chemically, why is it dangerous to mix alcohol with some other substances, including over-the-counter drugs? Which combinations are the most dangerous?

_____

_____

_____

_____

_____

_____

_____

_____

_____

_____

_____

_____

_____

_____

_____

_____

_____

_____

_____

_____

_____

# Chapter 5 -- Learning

**CHAPTER SUMMARY**

1. A definition of learning has to take into consideration the difference between learning and performance. Our definition makes this distinction by defining learning as the process by which experience or practice results in a relatively permanent change in what one is capable of doing. Ivan Pavlov developed classical conditioning that occurs when an originally neutral stimulus comes to elicit a response that was originally given to another stimulus. This occurs when the neutral stimulus is repeatedly paired with the other stimulus. B. F. Skinner expanded on the work of Thorndike and Watson to develop the theory of operant conditioning. Learning also includes imitation and cognitive learning.

2. In classical conditioning an unconditioned stimulus (US) is paired with a conditioned stimulus (CS) to produce a conditioned response (CR). A number of factors help in establishing conditioned responses. The CS must be strong and distinctive. The CS should be presented shortly before the US. The time lapse between the CS and US should be brief. There should be repeated pairings of the CS and US, and the pairings must be spaced properly. Extinction occurs if the CS is repeatedly presented alone. But in spontaneous recovery, the conditioning effect can appear again without any further pairings of the CS and US. Stimulus generalization occurs when responses are made to stimuli that are similar to, but not the same as, the original CS. Discrimination occurs when subjects learn to respond only to certain stimuli, but not to other, similar stimuli.

3. Operant conditioning occurs when desired responses are rewarded or reinforced and undesired responses are ignored or punished. Reinforcement after a response increases the likelihood that the response will be repeated. When a positive event serves as a reinforcer, it is called positive reinforcement. Negative reinforcement involves the removal of a stimulus after a response. It is used in escape training and avoidance training. Learned helplessness occurs when exposure to unavoidable, unpleasant events leads to an inability to avoid the unpleasant event even when avoidance is possible. Primary reinforcers or secondary reinforcers can be used in operant conditioning. Schedules of reinforcement can be either continuous (rewarding every response) or partial (rewarding only some responses), based on reinforcement of a certain number of responses or reinforcement after a certain interval of time. There are four basic schedules: fixed interval, variable-interval, fixed ratio, and variable ratio. Each type affects the rate of responding in different ways.

Shaping is a powerful procedure in which a part of behavior is reinforced, then the next parts are reinforced in successive steps until the whole behavior is performed. Extinction, spontaneous recovery, generalization and discrimination also occur in operant conditioning. Punishment, a stimulus that decreases the likelihood of a response when it is added to a environment, is used to change behavior. When dealing with children, adults should avoid reinforcing undesirable behaviors, immediately reward desirable behaviors, and supplement punishments with rewards for desirable responses. Preparedness is also a factor in operant conditioning.

4. Through insights we discover relationships that lead to problem-solving. We use cognitive maps to help us remember. Cognitive structures are learned through latent learning.

5.      Synapses are active processing units that play a key role in the physical representation of learning.  Learning in neural network models is represented by changing connection strengths.  The connection strength between two nodes increases when the nodes are activated together, and the magnitude of the increase is greater when the original strength is low.


## KEY TERMS AND CONCEPTS

History and Scope of the Study of Learning
> **Learning** is the process by which experience or practice results in a relatively
>> permanent change in what one is capable of doing
> Many psychologists define learning as a permanent change through experience
>> This ignores difference between what we can do and what we in fact do
> Latent learning - learning that is manifested some time later
>> It recognizes difference between learning and performance
> Pavlov revolutionized the study of learning with **classical conditioning**
>> Restraint was a limitation of classical conditioning
>> Operant conditioning relieved this limitation
> Skinner translated Thorndike's **law of effect** into **operant conditioning**


Classical Conditioning (Pavlov)
> **Unconditioned stimulus (US)** and **Unconditioned response (UR)**
> **Conditioned stimulus (CS)** and **Conditioned response (CR)**
> Important factors in classical conditioning
>> Establishing a response (five factors)
>>> CS must be strong and distinctive
>>> Order makes a difference
>>> Time lapse is important
>>> Repeated pairings are necessary
>>> Rate of strength depends on spaces of pairings
>> Extinguishing a response
>>> **Extinction** is decrease in response when CS is presented alone
>>> **Spontaneous recovery** is the sudden reappearance of a response
> Generalizing conditioned responses is called **stimulus generalization**
> Discrimination between stimuli is shown
> Higher-order conditioning is the basis of new learning on old learning


Operant Conditioning (Skinner)
> Reinforcement increases the likelihood of a response
>> Positive - positive event added to environment
>> Negative - negative event added to environment
>>> Used in **escape training** and **avoidance training**
>>>> **Learned helplessness** inability to avoid unpleasant events
>> Primary Reinforcers - reinforcing in and of themselves
>> Secondary Reinforcers - reinforcing only when paired with primary
> Schedules of reinforcement
>> Continuous reinforcement
>> Partial reinforcement
>>> Four common schedules
>>>> Fixed-interval schedules
>>>> Variable-interval schedules
>>>> Fixed-ratio schedules
>>>> Variable-ratio schedules

Encouraging a First Operant Response - getting that first response
    Hurry up and wait method
    Increase motivation
    Modeling
    Verbal instructions in personnel management
    Shaping
Extinguishing Operant Behaviors
    Extinction - withholding reinforcement
        Difficulty determined by four factors
Generalization and Discriminations
    **Stimulus generalizations** - same response to similar stimuli
    **Response generalizations** - different, but similar responses to stimulus
    **Stimulus discrimination** - different responses to different stimuli
Preparedness - animals more prepared to make certain responses than others
    Evidence favors this between and within species
    Preparedness also applies to humans

Cognitive Learning Theory
    **Insight** is the discovery of relations that lead to a solution of a problem
        Wolfgang Kohler was first to demonstrate insight
        Insights are common in human learning
    Latent Learning and Cognitive Maps
        Cognitive maps are mental pictures of our environment
    Cognition in Classical Conditioning
        Attention plays a role in classical conditioning
        Important for one stimulus to follow another
    Cognition in Operant Conditioning
        Important for reinforcement to follow a response

Psychobiological Analysis of Learning and Neural Networks
    Synapses are active processing units that play a key role in learning
    Learning in network models is represented by changing connection strengths

## DISCUSSION QUESTIONS AND EXERCISES

**Note: These questions and exercises ARE the learning objectives for this chapter. Answer them accurately in your own words and you will have mastered the most important material. We guarantee it.**

1.    <u>History</u> <u>and</u> <u>Scope</u> <u>of</u> <u>the</u> <u>Study</u> <u>of</u> <u>Learning</u>

a.    How is **learning** defined by your textbook? What problems occur with the definition that many psychologists give?

b.      What is **latent learning** and what does this concept recognize?

c.      What important discovery did Pavlov make?  How did he come upon this discovery?  Describe the major limitation of Pavlov's work.

d.      How does Thorndike explain his observations of the decrease in escape time as evidence of learning?  How did the work of Pavlov and Thorndike help experimental psychologists?

e.     The major milestone for behaviorism was when B.F. Skinner translated the law of effect into the theory of **operant conditioning**.  Define operant conditioning and identify what it allowed Skinner to accomplish.

2.     <u>Classical Conditioning</u>

a.     Characterize and give an *original* example of each of the following basic elements of classical conditioning:

   (1)     Unconditioned stimulus (US)

   (2)     Unconditioned response (UR)

   (3)     Conditioned stimulus (CS)

   (4)     Conditioned Response (CR)

b.    Provide an original example of classical conditioning.  Identify all of the relevant stimuli, all of the relevant responses, and the relationship between the stimuli and the responses.  Be creative!

c.    The likelihood of establishing a conditioned response depends on what five factors?

d.    Define both **extinction** and **spontaneous recovery**.  Give an example of each of these.  How does each of these occur?

e.  What term does the experiment with Little Albert exemplify?  How else is this term being used to help people?

f.  What would happen if stimulus generalization was carried too far?  Give an example.

g.  How did Pavlov demonstrate higher-order conditioning?  What did he speculate about learning?

3.  Operant Conditioning

a.  Contrast **positive** and **negative** reinforcement.  Use examples to illustrate the difference between the two procedures.

b.      Design an operant conditioning program in which you teach a dog to press a lever to turn off a light. Identify all of the relevant stimuli, responses, and reinforcers. (Note: you can use a regular light or a Bud Light.)

c.      What research has been done on **learned helplessness**? What explanations are given for why this may occur? Where may it be demonstrated in humans?

d.      What are **primary reinforcers**? How do **secondary reinforcers** rely on primary reinforcers? Give a brief example.

e.    Describe the four schedules of reinforcement and contrast the
      **partial-reinforcement** and **continuous** schedules of reinforcement.

f.    Identify some of the ways you would encourage the first operant response.  In
      contrast identify some of the ways you would extinguish the operant behaviors.

g.    What four factors affect how difficult it is to extinguish operant behaviors?
      Define **spontaneous recovery**.

h.    How are **punishment** and culture related?

i.    Give an original example of **stimulus generalization** and **response generalization**.  What is the opposite of stimulus generalization?

j.    The theory of **preparedness** would explain why rats jump to avoid foot shocks faster than pressing a bar because they are predisposed to jump when frightened. How would this theory apply to humans?

4. <u>Cognitive</u> <u>Learning</u> <u>Theory</u>

a. Wolfgang Kohler, a founder of Gestalt psychology, was among the first to demonstrate the importance of **insight** in problem solving. What is insight? How did he demonstrate it?

b. Discuss some of the implications of research on **cognitive maps**.

5. <u>Psychobiological</u> <u>Analysis</u> <u>of</u> <u>Learning</u> <u>and</u> <u>Neural</u> <u>Networks</u>

a. What important contribution have Kandel and his coworkers made concerning the neural representation of learning?

## POSTTEST

1.  The dependent variable in a classical conditioning study is the:
    a. conditioned stimulus
    b. number of conditioning trials
    c. strength of the conditioned response
    d. knowing whether or not the UCS is presented

2.  In classical conditioning, a _____ is reinforced; in operant conditioning, a _____ is reinforced.
    a. response; response
    b. response; stimulus
    c. stimulus; response
    d. stimulus; stimulus

3.  The process by which experiences result in a relatively permanent change in what one is capable of doing is called:
    a. intelligence
    b. knowledge
    c. learning
    d. performance

4.  A child discovers that her friends stop bothering him if he just ignores them. This is an example of:
    a. acquisition
    b. extinction
    c. shaping
    d. stimulus deprivation

5.  Pavlov found that meat placed on a dog's tongue will make the dog salivate. In Pavlov's terms, the meat is a(n):
    a. conditioned response
    b. conditioned stimulus
    c. unconditioned response
    d. unconditioned stimulus

6.  Identify who formulated the law of effect?
    a. Pavlov
    b. Skinner
    c. Thorndike
    d. Watson

7.  Shawn takes a course in which he is tested every two weeks. His studying drops right after a test, followed by a gradual increase to a rapid rate of studying as the next test approaches. His studying conforms to the pattern of responding maintained on _____ schedules.
    a. fixed-interval
    b. fixed-ratio
    c. variable-interval
    d. variable-ratio

8.   {T F} Spontaneous recovery is said to have happened when an animal emits a conditioned response from the animal's existing knowledge.

9.   The idea that animals are more prepared to make certain responses than others, and may find other responses more difficult to make is called

_____.

10.  _____ is to operant conditioning as _____ is to classical conditioning.
     a.   Wundt; Pavlov
     b.   Pavlov; Thorndike
     c.   Rescorla; Skinner
     d.   Skinner; Pavlov

11.  Cognitive learning theories attempt to explain how learning occurs using:
     a.   classical conditioning processes
     b.   classical, operant, and observational processes
     c.   observation and imitation
     d.   unobservable mental processes

12.  Which of the following is the BEST example of a secondary reinforcer?
     a.   clothing
     b.   food pellets
     c.   money
     d.   shelter

13.  The processing elements in neural networks are called _____. They enable the computation of probabilities or predictions.
     a.   connectors
     b.   models
     c.   neurons
     d.   nodes

14.  Positive reinforcement _____ the rate of responding; negative reinforcement _____ the rate of responding.
     a.   decreases; decreases
     b.   decreases; increases
     c.   increases; decreases
     d.   increases; increases

15.  After owning a car with a manual transmission, Bob buys a car with an automatic transmission. When first driving his new car, he keeps reaching for the nonexistent clutch and gear shift. This is an example of:
     a.   shaping
     b.   stimulus discrimination
     c.   stimulus generalization
     d.   stimulus response

16.  The studies on latent learning indicate that reinforcement is:
     a.   necessary for both learning and performance
     b.   necessary for learning but not for performance
     c.   necessary for performance but not learning
     d.   not necessary for either learning or performance

17.  _____ believes that the response patterns that are typical of the various schedules of reinforcement are a result of cognitive rules relating to the _____ between the responses and the likelihood of the consequence.
    a.   Kamin; contingencies
    b.   Kamin; expectations
    c.   Rescorla; contingencies
    d.   Rescorla; expectations

18.  Of the methods given to you in the textbook which of the following is NOT one of the ways to encourage a first operant response?
    a.   hurry up and wait
    b.   ignore the response
    c.   increasing motivation
    d.   modeling

19.  A person in extreme pain gets a shot of a painkilling drug and the pain goes away quickly.  Now, the person requests an injection whenever the pain occurs.  Which operant process does the example illustrate?
    a.   extinction
    b.   negative reinforcement
    c.   positive reinforcement
    d.   punishment

20.  "After a response, no consequence is presented or removed, and the probability of that behavior occurring is weakened/decreased."  This defines which of the following processes?
    a.   extinction
    b.   negative reinforcement
    c.   positive reinforcement
    d.   punishment

21.  Heather works in a dress factory where she earns $8 for each three dresses she hems.  Heather is paid on a _____ schedule.
    a.   fixed-interval
    b.   fixed-ratio
    c.   variable-interval
    d.   variable- ratio

22.  The sudden and irreversible learning of a solution to a problem is called _____ learning.

23.  Which of the following statements about learned helplessness is accurate?
    a.   caused by unavoidable, uncontrollable aversive stimulation
    b.   negative reinforcement causes it
    c.   positive reinforcement causes it
    d.   punishment is involved

24.  Classical conditioning could account for how a child learns to:
    a.   fear the dark
    b.   play football
    c.   talk
    d.   walk

25.  {T  F} Food, warmth, and shelter are examples of secondary reinforcers.

26. Assuming the reinforcer is the sound of the rattle, a baby's response of shaking a rattle is reinforced according to which type of schedule?
    a. fixed-interval
    b. partial reinforcement
    c. variable-interval
    d. continuous reinforcement

27. An unconditioned response is one that will occur:
    a. only after a number of conditioning trials
    b. only after the presentation of the conditioned stimulus
    c. without a stimulus
    d. without previous conditioning

28. The blocking effect in classical conditioning is important because it illustrates that:
    a. mere contiguity is not enough to produce an association between stimuli
    b. one stimulus need not signal the occurrence of another for an association to be formed
    c. unconditioned stimuli can block conditioned stimuli
    d. unpleasant stimuli, like shock, can block pleasant stimuli

29. Given the same frequency of reinforcement, _____ schedules generate higher rates of responding than do _____ schedules.
    a. fixed; variable
    b. interval; ratio
    c. ratio; interval
    d. variable; fixed

30. In classical conditioning, the subject learns to respond to one CS but not to a different CS. This is an example of:
    a. conditioned forgetting
    b. extinction
    c. stimulus discrimination
    d. stimulus generalization

31. Which of the following is an example of a variable-interval schedule?
    a. A gambler wins on 8 out of 10 poker hands on the average
    b. A bird gets food the first time it pecks a key after 15 sec. have passed
    c. A fisherman gets a bite once every 10 minutes on the average
    d. Bobby gets one chocolate-chip cookie for every 12 push-ups he does

32. Operant conditioning assumes that:
    a. events that follow behavior affect whether the behavior is repeated
    b. one's mental processes (e.g., memory and perception) mediate what behaviors one does in a situation
    c. one learns by watching others' behaviors
    d. voluntary behaviors are reflexive

33. {T  F} Tolman's work demonstrates that humans can develop cognitive maps, but rats cannot.

34. The process of selectively reinforcing responses that are closer and closer approximations of some desired response is called:
   a. selection
   b. shaping
   c. step-wise conditioning
   d. stimulus discrimination

35. As a child, whenever Cindy had a tantrum her parents tried to quiet her down. When she turned 9, though, her parents stopped attending to her when she began to scream and cry. Six months later, Cindy no longer exhibits tantrums. The parents' attention at the beginning of the example was:
   a. a negative reinforcer
   b. a positive reinforcer
   c. a punisher
   d. an extinguisher

36. The steady, rapid responding of a person playing a slot machine is an example of the pattern of responding typically generated on a _____ schedule.
   a. fixed-interval
   b. fixed-ratio
   c. variable-interval
   d. variable-ratio

37. Which of the following is NOT one of the important factors in classical conditioning?
   a. there must be repeated pairing of the CS and US
   b. the CS must be strong and distinctive
   c. the first pairing must be at a slow rate
   d. the order in which the CS and US are presented is important

38. If her daddy comes home smiling and gives her a kiss, eight-year-old Molly asks him for money for her piggy bank because she knows she will get a dollar then--but not at any other time. The father's smile and kiss is a/an _____ stimulus.
   a. conditioned
   b. discriminative
   c. higher-order
   d. unconditioned

39. In a Skinner box, the dependent variable was:
   a. the force with which the lever is pressed
   b. the rate of responding
   c. the schedule of reinforcement used
   d. the speed of the cumulative recorder

40. The phenomenon of higher-order conditioning shows that:
   a. auditory stimuli are easier to condition than visual stimuli
   b. an already established CS can be used in the place of a natural UCS
   c. only a genuine, natural UCS can be used to establish a CR
   d. visual stimuli are easier to condition than auditory stimuli

41. According to Thorndike, the presence of food on the outside of a cat's puzzle box:
    a. created a mental expectation of food in the cat's mind
    b. increased the exploratory drive of the cat, leading to eventual escape
    c. strengthened the association between the puzzle box and the responses that led to escape from the box
    d. decreased the exploratory drive of the cat, leading to eventual learned helplessness

42. Of the following phenomena, which one best explains the spreading of phobias to objects similar to the one to which the phobia was originally acquired?
    a. discrimination
    b. extinction
    c. generalization
    d. spontaneous recovery

43. Sam has his driving privileges temporarily suspended after getting a speeding ticket. This is an example of:
    a. extinction
    b. negative punishment
    c. positive reinforcement
    d. punishment

44. _____ training occurs when a subject learns to make a response to avoid something unpleasant.

45. In general, the longer the delay between a response and reinforcement:
    a. the faster conditioning proceeds
    b. the less effective the reinforcer becomes
    c. the more effective the reinforcer becomes
    d. the more slowly conditioning proceeds

46. Many psychologists define learning as a permanent change in behavior through experience. The primary problem with this definition is that
    a. It does not take into account the difference between learning and performance.
    b. Some things are learned through maturation and not experience.
    c. Some things may be learned only temporarily.
    d. It focuses too much on behavior and not enough on emotions.

47. Which of the following statements is TRUE?
    a. Operant behaviors decrease immediately when reinforcement is withheld.
    b. Operant behaviors increase for a brief period after reinforcement is withheld.
    c. Operant behaviors increase for an extended amount of time after reinforcement is withheld.
    d. Operant behaviors remain steady after reinforcement is withheld.

48. In one experiment, researchers discovered that younger rats forgot conditioned responses more rapidly than younger ones. This experiment provides evidence for:
   a. differences in stimulus generalization as a function of age
   b. differences in response generalization as a function of age
   c. differences in preparedness as a function of age
   d. differences in preparedness between species

49. An important difference between the cats in Thorndike's puzzle box experiment and the chimpanzees in Kohler's experiment is:
   a. the chimpanzees solved their problem through trial and error whereas the cats solved their problem by insight
   b. the cats solved their problem through trial and error whereas the chimpanzees solved their problem by insight
   c. the cats were trained through operant conditioning whereas the chimpanzees were trained through classical conditioning
   d. the chimpanzees were trained through operant conditioning whereas the cats were trained through classical conditioning

50. The concept of "blocking" is important because it provides supporting evidence to those who believe that
   a. performance does not necessarily reflect learning strength
   b. classical conditioning is a strictly behavioral strength
   c. learning does not always show itself immediately in performance
   d. cognitive processes play a role in classical conditioning

## POSTTEST ANSWERS

1. c
2. c
3. c
4. b
5. d
6. c
7. a
8. T
9. preparedness
10. d
11. d
12. c
13. d
14. d
15. c
16. c
17. c
18. b
19. b
20. a
21. b
22. insight
23. a
24. a
25. F
26. d
27. d
28. a
29. c
30. d
31. c
32. a
33. F
34. b
35. b
36. d
37. c
38. b
39. b
40. b
41. c
42. c
43. d
44. Avoidance
45. d
46. a
47. b
48. c
49. b
50. d

# PSYCH JOURNAL

**Please use the following pages to record your thoughts and feelings about the following questions.**

1.    Helen Keller celebrated the day that Anne Sullivan came to teach her as her "soul's birthday." This chapter showed that Sullivan was an effective teacher, in part because she followed fundamental principles of learning. Has any teacher had a major impact on your life? Can you remember times when your teachers or you used the learning principles that are presented in this book?

_____

_____

_____

_____

_____

_____

_____

_____

_____

_____

_____

_____

_____

_____

_____

_____

_____

_____

_____

2. Have you ever seen a small child throw a tantrum in a store because he wants something? Did you feel at the time that the parents should "do something" so the kid wouldn't annoy other people? Why might the parents have taken the approach they did? What would you suggest they do? Why is parents' consistency of response so important?

_____

_____

_____

_____

_____

_____

_____

_____

_____

_____

_____

_____

_____

_____

_____

_____

_____

_____

_____

_____

3. Most of us have read or heard about studies developed to measure conditioning and learning in animals that have sometimes used somewhat harsh methods. Do you think such methods should be used? How might they be modified?

_____

_____

_____

_____

_____

_____

_____

_____

_____

_____

_____

_____

_____

_____

_____

_____

_____

_____

_____

_____

# Chapter 6 -- Memory and Cognition

**CHAPTER SUMMARY**

1.      The Greeks and Romans studied memory as an art.  Herman Ebbinghaus devised the first experiments in memory.  The behaviorists used Ebbinghaus's experiments to describe memory in terms of stimuli and responses.  By the 1960s cognitive psychologists began to infer unobservable memory processes from observable performance.  Theories of multiple memories and levels of processing have emerged to explain memory.

2.      Explicit memory is tested by recall tests and recognition tests.  Recall tests include fill-in-the-blank and essay questions.  Recognition tests include multiple-choice and matching questions.  One way to measure implicit memory is by using a priming test.

3.      Sensory memory holds sensory impressions for only 1 or 2 seconds.  Both visual and auditory sensory memory can be affected by backward masking.  Short-term memory holds information transferred from sensory memory.  Memory will last only 30 seconds unless it is repeated.  Maintenance rehearsal improves short-term memory.  Long-term memory holds information that is transferred from short-term memory through rehearsal or some other process.  We tend to recall items better when they are presented at the beginning or end of a list (serial-position effect).  We also store memory aspects of our experience such as the voices of old friends.

4.      Through attention we notice stimuli and ignore irrelevant stimuli.  We use verbal codes for verbal materials such as digits, letters, and words.  We seem to refer visual codes for material that is difficult to describe.  The "magical number" $7 \pm 2$ represents the number of items in the memory span of most normal adults.  Multiple memory theorists say this is due to limited storage space in short-term memory.  Chunking information improves short-term memory.

5.      Emotionally charged events may cause us to have vivid flashbulb memories.  Other times these memories may lose accuracy.  We tend to remember meaning as opposed to the exact words used to convey ideas.

6.      We store our knowledge in three kinds of long-term memory; episodic memory, semantic memory, and procedural memory.  We use procedural knowledge to put our other knowledge to use.  Connectionist and parallel distributed processing models consist of networks of episodic, semantic and procedural memory that enable implicit and explicit memory to work together.  We use semantic memory to organize episodic memory.  The organization may reflect a hierarchy of categories.  It is difficult to add new associations to semantic memory.  Culture and gender can also influence semantic memory.  Reconstruction of long-term memory occurs often.  Eyewitness testimony may be influenced by new information and leading questions.

7.      Retrieval from short-term memory involves a search process that is not open to conscious inspection.  Retrieval from long-term memory involves retrieval cues.  The tip-of-the-tongue phenomenon reveals that we have information in our memory that may be hard to find.  Anxiety, costs, and rewards may influence our ability to remember or to verbalize our memories.

8.      Levels-of-processing theorists reject the distinction between short-term memory and long-term memory (they recognize sensory memory). Instead, they say that memory is a unified system that involves surface and deep processing. We forget some things rapidly because we process them minimally. We remember other things longer because we process them more deeply. We can only remember the magical number 7 +- 2 because we have limited processing capacity. Elaborative rehearsal occurs as part of deeper processing. We think about the relationship between new information and old, thereby increasing our ability to remember.

9.      Trace decay may cause memory loss. Traces start out strong, and their strength is maintained through use. But traces weaken and fade away if they are not used. Interference theory holds that we forget information because other information gets in the way. Proactive interference means that something already learned interferes with our ability to learn something new. Retroactive interference means that new learning interferes with what we already know. Some theorists believe that new memories may dissolve old memories; others argue that they only make the older memories harder to find.

10.      Neurobiological changes that underlie short-term and long-term effects of learning and memory include changes in the excitability of cell membranes and the establishment of new connections between neurons. Damage to the brain has resulted in memory loss. The hippocampus is important in explicit memory and in forming "new" long-term memories. Alzheimer's disease is associated with a degeneration of neurons that manufacture acetylcholine. Electrical and chemical stimulation of the brain can enhance memory.

11.      Some methods you can use to improve your memory. Make sure that you are paying attention. Remove all distractors. Organize the material you want to remember. Use mnemonic systems such as the keyword method and verbal mediation. Use organization techniques when you review. To retrieve information, search your memory systematically and try to recall the context in which the memory was formed.


**KEY TERMS AND CONCEPTS**

History and Scope of the Study of Memory
    Memory is the system that allows us to retain information over time
    Greeks and Romans studied memory as art first
    Herman Ebbinghaus devised the first experiments in memory
        Used himself as a subject
        Used unrelated, three-letter, non-sense syllables
        Behaviorists used this to describe memory
    By 1960s cognitive psychologists inferred unobservable memory from
        performance
    Theories of multiple memories and levels of processing have emerged

Measuring Explicit and Implicit Memory Processes
    Schema - organized set of knowledge about concepts
    Script - organized knowledge set about how events are sequenced
    Mental Model - organized set of knowledge about how things work

Explicit memory - tested by recall tests and recognition tests
    Recall tests include fill-in-the-blank and essay questions
        Measure a person's ability to retrieve information from memory
        Cued recall
        Free recall
    Recognition tests include multiple-choice and matching questions
        Usually easier than recall tests
Implicit memory is measure by a priming test
    Influences behavior or thought without entering consciousness
    Related to implicit learning

An Information Processing Model with Multiple Memories
    Multiple memory theory
    Sensory memory holds sensory impressions
        Only for 1 or 2 seconds
        Backward masking affects both visual and auditory sensory memory
    Short-term memory holds information transferred from sensory memory
        Will last only 30 seconds unless it is repeated
        Maintenance rehearsal improves short-term memory
    Long-term memory holds information transferred from short-term memory
        This is done through rehearsal or other processes
        Serial-position effect
            Items at beginning and end of lists are recalled better
                Primacy effect - recall from long-term
                Recency effect - recall from short-term
        Sensory aspects of our experience are stored in long-term memory
            Voices of old friends
            Complex movements
            Smells

Encoding in Short-Term Memory
    Encoding
        Collection of memory processes that select incoming stimuli
    Attention helps ignore irrelevant stimuli and notice important stimuli
        Dichotic listening
    Verbal codes are used for verbal materials
        Include digits, letters, and words
    Visual codes are for material that is difficult to describe
    Memory span - read through and recall with no mistakes
        The "magical number"
            Normal adult remembers $7 \pm 2$ items in the memory span
            Multiple memory theorists say it is due to limited storage
    Chunking is grouping stimuli into meaningful units
        This improves short-term memory

Encoding in Long-Term Memory
    Emotionally charged events cause us to have vivid memories
        Flashbulb memories are vivid memories of important events
        Other times they may lose accuracy
    We remember meanings as opposed to exact words to convey ideas

Retention in Long-Term Memory
    Three kinds of long-term memory
        Episodic memory - memories about events
        Semantic memory - our background knowledge about words, and rules
        Procedural memory - our background knowledge about how to do things
            Procedural knowledge puts our other knowledge to work
            Much of procedural knowledge is in implicit memory
    Connectionist and parallel distributed processing models
        These consist of networks of episodic, semantic and procedural memory
            Implicit and explicit memory work together here
    Semantic memory is used to organize episodic memory
        Free association tests
            Reflect a hierarchy of categories
        Difficult to add new associations
        Culture and gender can influence semantic memory
    Reconstruction of long-term memory occurs often
        Eyewitness testimony may be influenced by reconstruction

Retrieval - getting information out of memory
    Short-term retrieval involves a search process
        This is not open to conscious inspection
    Long-term retrieval involves retrieval cues
        Tip-of-the-tongue phenomenon
            Information that is hard to find
        Anxiety, costs, and rewards influence our ability to remember memories

Level of Processing
    These theorists reject distinction between short and long-term memory
        They do recognize sensory memory
        The rest of memory is a unified system
            Involves surface and deep processing
            Forgetting occurs rapidly because of minimal processing
            Remembering occurs because of deep processing
    We can only remember $7 \pm 2$ because of limited processing capacity
    Elaborative rehearsal
        New and old information increases our ability to remember

Forgetting is an inability to remember
    Trace decay may cause memory loss
        Traces maintain strong if used but weaken and fade if not used
    Interference theory
        Other information conflicts with current information
        Proactive interference - old interferes with new
        Retroactive interference - new interferes with old

The Neurobiological Basis of Learning and Memory
    Neurobiological changes underlie short-term and long-term effects of learning
        Excitability of cell membranes
        Establishment of new connections between neurons
    Damage to brain results in memory loss
        Hippocampus is important in learning and memory
            Amnesia - partial forgetting of past events
                Retrograde amnesia - loss of events before injury
                Anterograde amnesia - loss of events after injury

Korsakoff's syndrome - thiamine deficiency (apathy, confusion)
Alzheimer's disease is associated with low acetylcholine
Electrical and chemical stimulation of brain enhances memory

Improving Your Memory
Pay attention and remove distractors
Organize material you want to remember
Use mnemonic systems
Keyword method and verbal mediation
Use organization techniques when you review
To retrieve information, search your memory systematically
Try to recall the context in which the memory was formed

## DISCUSSION QUESTIONS AND EXERCISES

**Note: These questions and exercises ARE the learning objectives for this chapter. Answer them accurately in your own words and you will have mastered the most important material. We guarantee it.**

1.  The History and Scope of the Study of Memory

a.  Define **memory** and describe the research done by Ebbinghaus, including one of his findings.

b.  How did behaviorists use Ebbinghaus's methods in their investigations?

2.  Measuring Explicit and Implicit Memory Processes

a.  Define each of the following terms:

    (1)  **Schema**

    (2)  **Script**

    (3)  **Mental Model**

b.  What methods are used to study both **explicit memory** and **implicit memory**? Contrast **recall** tests and **recognition** tests.

3.  An Information Processing Model with Multiple Memories

a.  What are the three types of memory? Give a brief description of their characteristics.

b.	Sperling demonstrated some very important characteristics of sensory memory in experiments.  What are these important findings?

c.	Identify the term that describes the process of repeating information in order to retain it in short-term memory.  Give an *original* example.

d.	Describe two ways that short-term and long-term memory differ.  What is the **serial-position** effect in free recall?

e.	What determines the **recency** and **primacy** effects?  Give examples.

f.  What is **encoding**? Without the ability to pay **attention** to certain things our minds would surely drown in an ocean of stimuli. How has research on **dichotic listening** helped to understand attention?

g.  When would we prefer to use verbal codes? When would we use visual codes?

h.  What is the "Magical Number"? Describe the process in which we can expand short-term memory.

4.  Encoding in Long-Term Memory

a.  Define **flashbulb memories**. Do you have any vivid memories of sudden, emotionally charged events?

b.   What role does sentence meaning play in long-term memory?

5.   Retention in Long-Term Memory

a.   Characterize each of the following types of long-term memory:

   (1)   Episodic memory

   (2)   Semantic memory

   (3)   Procedural memory

b.   How do neural network models help us to understand explicit and implicit memory?

c. What evidence about the interaction between semantic and episodic memory has been shown in research using **free association** tests.

d. What is meant by reconstruction in memory? Give an example.

6. <u>Retrieval</u>

a. How does retrieval from short-term and long-term memory differ? Identify some of the cues that are utilized to aid in retrieval from long-term memory.

b. How can costs and rewards influence the testimony of witnesses in a trial? How does Buckhout use **signal-detection** theory?

7.   <u>Levels of Processing</u>

a.   How do levels-of-processing theorists view memory?  Identify some of the differences between levels-of-processing theorists and multiple memory theorists.

b.   Why does deeper processing make memory longer?  What evidence is there to support this?

c.   What term describes the thinking about the relationship between new information and old information that is already in memory?  How does this contrast with surface-level processing?

8.   <u>Forgetting</u>

a.   Define **forgetting**.  How does **trace decay theory** explain forgetting?

b.      Give one example of forgetting by retroactive interference and one example of forgetting by proactive interference.

9.      The Neurobiological Basis of Learning and Memory

a.      What effect does learning have on our neurobiological processes?  How have computers helped in this field?

b.      What methods are used by researchers to find the location of **memory traces**?

c.      What is the difference between retrograde and anterograde amnesia?

d.  Briefly describe Korsakoff's syndrome and Alzheimer's disease.  What are some of the symptoms and known causes of these diseases?

10.  Improving Your Memory

a.  Identify some of the methods described in your textbook to improve your memory.

**POSTTEST**

1.  The multiple memory theory states that:
    a.  individuals have many kinds of memory with different layers or depths
    b.  individuals have five different types of memory
    c.  individuals have three types of memory: sensory, short-term and long-term
    d.  information is processed into abstract forms as it is processed deeper

2.  Your memory of how to tie your shoe is contained in your _____ memory.
    a.  declarative
    b.  episodic
    c.  procedural
    d.  semantic

3.  The system that allows us to retain information over time is called _____.

4.     If the three memory stores were ordered in terms of levels of processing, which of the following sequences would be the correct order for progressively deeper processing?
    a.     long-term memory, short-term memory, sensory memory
    b.     sensory memory, long-term memory, short-term memory
    c.     sensory memory, short-term memory, long-term memory
    d.     short-term memory, sensory memory, long-term memory

5.     _____ involves focusing awareness on a narrow range of stimuli or events.
    a.     Attention
    b.     Clustering
    c.     Elaboration
    d.     Encoding

6.     A primacy effect probably occurs because the first term in a list:
    a.     is paid less attention to than later items
    b.     is still available in short-term storage
    c.     can no longer be retrieved from sensory storage
    d.     has had the most opportunity for rehearsal

7.     Remembering how to fly a kite is _____ memory; remembering who helped you learn is _____ memory; remembering where you learned is _____ memory.
    a.     episodic; semantic; sensory
    b.     procedural; episodic; episodic
    c.     procedural; semantic; episodic
    d.     semantic; semantic; episodic

8.     If you associate a concrete word with a to-be remembered abstract word and then generate an image to represent the concrete word, you are using:
    a.     a semantic network
    b.     an acrostic
    c.     the link method
    d.     the keyword method

9.     When you listen to a lecture, the information is held in _____ memory until you write it in your notes.
    a.     long-term
    b.     sensory
    c.     short-term
    d.     trace

10.     Eyewitnesses make errors in retrieval because they sometimes do all of the following except:
    a.     fill in memory gaps with inferences
    b.     include elements of both the original event and subsequent events in their recollection
    c.     include information in the questions about the event that can be incorporated into the recollection
    d.     suffer from low anxiety, which hinders retrieval

11. This multiple-choice question is an example of a _____ measure of retention.
    a. recall
    b. recognition
    c. reiteration
    d. relearning

12. Elaborative rehearsal is to _____ storage as maintenance rehearsal is to _____ storage.
    a. short-term; long-term
    b. sensory; short-term
    c. long-term; short-term
    d. short-term; sensory

13. Which of the following statements about sensory memory is FALSE?
    a. Information in sensory memory is stored in a highly processed form.
    b. It is the first component of the human memory system.
    c. It preserves information for a very brief time.
    d. There is a sensory memory system for each of the senses.

14. As you study for your next test, the material you are processing sometimes becomes more difficult to remember because of interference from older material you have learned. This is called _____.
    a. cue-dependent interference
    b. proactive interference
    c. retroactive interference
    d. trace decay

15. A conceptual hierarchy:
    a. is a multilevel classification system based on common properties
    b. is a way of organizing information in short-term memory
    c. is the same thing as a neural network
    d. is the same thing as a semantic network

16. The distinction between the levels of processing and multiple memory theories is:
    a. levels of processing theory says there is no sensory store
    b. multiple memory theory says there is no sensory store
    c. the levels of processing theory believes there are only two types of memory, while the multiple memory theory says there are three
    d. the levels of processing theory believes there are three types of memory while the multiple memory theory says there are only two

17. Walter is thrown from his bicycle and suffers a severe blow to the head, resulting in loss of memory for events that occurred before the accident. This is an example of:
    a. anterograde amnesia
    b. motivated forgetting
    c. retroactive interference
    d. retrograde amnesia

143

18. Liz memorized her shopping list. When she got to the store, however, she had forgotten many of the items from the middle of the list. This is an example of:
    a. inappropriate encoding
    b. proactive interference
    c. retrograde amnesia
    d. the serial-position effect

19. The process of getting information out of memory is called:
    a. encoding
    b. purging
    c. retention
    d. retrieval

20. Eric is reminiscing about the first car he owned in high school and how he felt the first time he drove it through town. This is an example of _____ memory.
    a. declarative
    b. episodic
    c. procedural
    d. semantic

21. {T F} Long-term memory is independent of short-term memory.

22. A fan's organized set of expectations about how a movie star is supposed to act is an example of a:
    a. chunk
    b. schema
    c. script
    d. semantic network

23. The capacity of short-term memory is:
    a. seven, plus or minus two numbers
    b. seven, plus or minus two letters
    c. seven, plus or minus two chunks
    d. not expandable at all

24. Flashbulb memories are:
    a. chronological recollections of personal experiences
    b. memories for actions, skills, and operations
    c. memories for factual information
    d. unusually vivid and detailed recollections of momentous events

25. A list of words is presented. Which would be an implicit test of the words?
    a. ask subjects to write down all the words they remember
    b. ask the subjects to relearn the words after a period of time
    c. give the first three letters of the words and ask subjects to complete the stems with any word that comes to mind
    d. present a long list of words, half new and half old; ask subjects to say whether each word is old or new

26. The Atkinson-Shiffrin model for memory is:
    a. a multiple memory theory
    b. a contradiction to Freudian theory
    c. an extension of B. F. Skinner's theory
    d. the only memory theory that exists

27. If you asked when your grandmother's birthday was, this is a _____ question.
    a. recall
    b. recognition
    c. reiteration
    d. relearning

28. The theory that holds that memory traces fade away in time is called _____ theory.

29. When Ebbinghaus discovered that one study trial was sufficient for remembering lists of 7 or 8 nonsense syllables, he had actually discovered the limit of what today we call _____ memory.
    a. long-term
    b. sensory
    c. short-term
    d. visual

30. The usefulness of the sensory store is limited by:
    a. its capacity
    b. its duration
    c. both a and b
    d. neither a nor b

31. The hippocampus seems to be essential for:
    a. maintaining one's balance
    b. proactive and retroactive inhibition
    c. the formation of new long-term memories
    d. the recall of old memories

32. Your knowledge that frogs hop, that the sun sets in the west, and that 2 - 2 = 0 is contained in your _____ memory.
    a. implicit
    b. procedural
    c. semantic
    d. structural

33. A student's organized set of expectations about how to study for and take an exam is a:
    a. cluster
    b. conceptual hierarchy
    c. semantic network
    d. script

34. {T F} One criticism of connectionist models of memory is that they have been formulated without consideration of the way the brain and nervous system work.

35. In anterograde amnesia:
    a. a person loses memories of events that occur after a head injury
    b. a person loses memories of events that occurred prior to a head injury
    c. new information impairs the retention of previously learned information
    d. previously learned information interferes with the retention of new information

36. The ability of a stimulus to wipe out the sensory memory of a preceding stimulus is called _____ _____.

37. The reason chess masters had a better short-term memory of chess moves than novice players was
   a.   they are more intelligent
   b.   they cheated
   c.   they had larger memory spans
   d.   they were able to chunk the moves into meaningful units

38. The amount of information that you can hold in short-term memory can be increased by:
   a.   practicing to expand capacity beyond seven chunks
   b.   reorganizing information into larger chunks
   c.   learning techniques to expand your capacity
   d.   sequencing information into groups

39. The cause of brain deterioration in patients with Korsakoff's syndrome is:
   a.   a prolonged Vitamin B1 (thiamine deficiency)
   b.   glucose deficiency in the brain
   c.   old age
   d.   the direct effects of alcohol on the brain

40. The tip-of-the-tongue phenomenon:
   a.   is a temporary inability to remember something you know
   b.   is clearly due to a failure in retrieval
   c.   both a and b
   d.   neither a nor b

41. A good strategy for minimizing interference with retention is to:
   a.   conduct a thorough review of material as close to exam time as possible
   b.   engage in massed practice
   c.   overlearn the material
   d.   spend less time on rote repetition of the material

42. To describe memory as a "reconstructive process" means:
   a.   it involves deletion of some portions or facts and the insertion of new portions or facts
   b.   it involves retrieving information from short-term memory instead of from long-term memory
   c.   it is an exact reproduction of the original stimulus
   d.   it is done totally without notes or other aids

43. A recency effect probably occurs because the final items in a list:
   a.   are paid more attention than earlier items
   b.   are still available in short-term storage
   c.   have had the least opportunity for rehearsal
   d.   have had the most opportunity for rehearsal

44. Mmemory cannot be directly _____ ; it must be _____ from performance.
   a.   assessed; measured
   b.   inferred; assessed
   c.   measured; observed
   d.   observed; inferred

45. You move to a new house and memorize your new phone number. Now, you can't remember your old phone number. This is an example of:
   a. motivated forgetting
   b. proactive interference
   c. retroactive interference
   d. retrograde amnesia

## POSTTEST ANSWERS

1.   c
2.   c
3.   memory
4.   c
5.   a
6.   d
7.   b
8.   d
9.   c
10.  d
11.  b
12.  c
13.  a
14.  b
15.  a
16.  c
17.  d
18.  c
19.  c
20.  b
21.  F
22.  b
23.  c
24.  d
25.  c
26.  a
27.  a
28.  trace delay
29.  c
30.  b
31.  c
32.  c
33.  d
34.  F
35.  a
36.  backward masking
37.  d
38.  b
39.  a
40.  c
41.  a
42.  a
43.  b
44.  c
45.  c

**PSYCH JOURNAL**

**Please use the following pages to record your thoughts and feelings about the following questions.**

1. Think about all the ways in which witnesses in the Sacco and Vanzetti case illustrated principles of memory processes in the human information processing system. Are any of these principles illustrated in your own experience? Were you ever a witness to a crime?

_____

_____

_____

_____

_____

_____

_____

_____

_____

_____

_____

_____

_____

_____

_____

_____

_____

_____

2. Do you have "flashbulb" memories, like remembering where you were and what you were doing when the Challenger space shuttle exploded? Describe one of these memories that you recall. Why do you think this event is so vivid in your mind?

_____

_____

_____

_____

_____

_____

_____

_____

_____

_____

_____

_____

_____

_____

_____

_____

_____

_____

_____

3.   It is helpful to you to relate abstract concepts to your own life experiences through writing? Can you think of a paper you wrote some time ago from which you still remember many ideas?

_____

_____

_____

_____

_____

_____

_____

_____

_____

_____

_____

_____

_____

_____

_____

_____

_____

_____

_____

_____

_____

4. Have you and a family member or a friend ever disagreed about how a particular event took place? How would you explain that disagreement?

_____

_____

_____

_____

_____

_____

_____

_____

_____

_____

_____

_____

_____

_____

_____

_____

_____

_____

_____

_____

5. Have you ever used mnemonic devices (like "every good boy deserves fudge" to remember musical scales) to improve your retention? Can you still recall that information because of the mnemonics?

_____

_____

_____

_____

_____

_____

_____

_____

_____

_____

_____

_____

_____

_____

_____

_____

_____

_____

_____

_____

_____

# Chapter 7 -- Language, Thought, and Intelligence

## CHAPTER SUMMARY

1.      The traditionally separated areas of language, thought, and intelligence are being drawn together to solve the mystery of artificial intelligence. In order to program computers to do what humans do, scientists must understand the interactions of language, thought, and intelligence. Cognitive science views humans as information processing systems combining elements of language, thought, and intelligence in many different ways.

2.      All human languages are (1) structured to enable creative usage, (2) interpersonal, and (3) meaningful or referential. Creative usage means that we can generate and interpret sentences that we have never encountered before. Language is interpersonal in that speakers or writers use language to communicate their thoughts to others. Language is meaningful because it typically expresses what we think about and know. Animal communication can be interpersonal and meaningful or referential. However, animal communication does not involve creative usage.

3.      According to the Sapir-Whorf hypothesis, people who speak different languages think and perceive the world differently. Negative attitudes could cause some people to create sexist language. This language could then cause other people to develop negative attitudes toward women. Through imaginal thought we manipulate mental images without the use of words. We form concepts based on similarities. Family-resemblance or natural concepts are categories that are often represented by a prototype. We use either a holistic approach or an attribute-by-attribute when we manipulate symbols to form natural concepts.

4.      Problem solving involves three steps: defining the problem, making plans, and making decisions. Defining a problem correctly is sometimes difficult when people experience functional fixedness. Planning often involves outlining a series of subgoals. Experts usually spend more time planning than novices. Making decisions includes evaluating and choosing the best approaches to problem solving. Trial and error involves trying different strategies until one works. Hypothesis testing occurs when people test an educated guess on the basis of available evidence. Some problems are solved through the use of algorithms and heuristics. Scientists are working on artificial intelligence so that computers can solve more problems.

5.      Achievement tests and aptitude tests are generally used to test intelligence. Most aptitude tests are standardized. They must be reliable and valid. The best known aptitude test is the IQ test that produces a score based on mental and chronological age. IQ scores may represent a range from mental retardation to mentally gifted. IQ may not represent the intellectual changes that take place with age. Researchers have found that crystallized intelligence increases with age while fluid intelligence decreases. It has been argued that IQ tests are not fair to African Americans, Hispanics, poor whites and other cultural subgroups. Investigators have used three procedures to study the interaction between genetics and environment in relation to intelligence. These are studies of identical twins who are raised apart; studies comparing identical and fraternal twins, and studies of adopted children.

The most heated debate about intelligence concerns the fact that in the United States African-American people average 10 to 15 points lower on IQ tests than white people. Although some have tried to use this data to argue that African-Americans are innately less intelligent than whites, it is clear that this conclusion is not true. Psychologists have countered this conclusion with many arguments including: Hereditary differences should be considered within groups not in comparing groups. IQ tests are biased against African-Americans and other minorities. The IQ gap between African-American and European-American children is narrowing as environmental conditions improve for African-Americans.

6.      Creativity involves divergent thinking. The Divergent Production Test, which presents open-ended questions, and the Remote Associates (RAT), which measure's a person's ability to see relationships between things that are remotely associated, are often used to assess creativity. Creative people are less conforming to social norms and they take more risks. External rewards tend to make people less creative. Creative people tend to draw on highly organized knowledge in their field of expertise.

7.      According to Gardner, our mental abilities include more than those that enable us to succeed in school. Gardner has proposed that there are six separate intelligences; linguistic, logic-mathematical, spatial, musical, bodily-kinesthetic, and personal. IQ tests do not presently tests all of these intelligences. Sternberg suggests that a comprehensive theory must cover three aspects of intelligence. The componential aspect basically represents what is covered by traditional IQ tests. The experiential aspect refers to the ability to combine experiences in insightful ways. And the contextual aspect emphasizes adaptation to the environment.

## KEY TERMS AND CONCEPTS

Scope and History
    Language, thought, and intelligence are traditionally separated
        Drawn together to solve mystery of artificial intelligence
    Cognitive science views humans as information processing systems
        Transform energy into symbols
        Manipulate, transform, and store the symbols
        Select and execute responses
        The systems combine language, thought, and intelligence
    Behaviorism had downfalls because of inability to explain thought and language

Language
    Human language is a means of communicating by vocal sounds, marks, gestures
    Characteristics of human languages
        Structured to enable creative usage
            Rules of grammar are principles for structuring language
            Generate and interpret sentences not used previously
                Letters - smallest units of meaningful language
                    Orthographic rules - legal and illegal letters
                Phonemes - smallest units of speech sound
                    Phonological rules - legal and illegal sounds
                Morphemes - combination of letters or phonemes

Interpersonal
        People use language to communicate thoughts to others
            Goal is to create similar thoughts in listeners
    Meaningful or referential
        Expresses what we think about and know
Animal communication can have some of these same characteristics
    Does not involve creative usage

Thought
    Thought is the mental activity of manipulating symbols
    Sapir-Whorf hypothesis
        Those who speak different languages think differently
    Sexist language caused by negative attitudes
        This could cause others to develop negative attitudes about women
    Imaginal thought is used to manipulate mental images without words
    Concepts are formed based on similarities
        Family-resemblance and natural concepts
            Represented by a prototype
            Two approaches we use to form natural concepts
                Holistic
                Attribute-by-attribute

Problem Solving
    Problem solving is thinking aimed at overcoming obstacles in path of a goal
    Three steps to problem solving
        Defining the problem
            Difficult when people experience functional fixedness
                Functional fixedness - people see objects in familiar use
        Making plans
            Outlining a series of subgoals
            Experts spend more time planning than novices
        Making decisions and evaluating strategies
            Trial and error until one works
            Hypothesis testing
                Testing of educated guess based on evidence
                    Confirmation bias
            Algorithms and heuristics are also used
                Representative heuristic
                Availability heuristic

Intelligence: Assessment, Controversies, and Theories
    Intelligence is the capacity to learn and use information
    Achievement tests and aptitude tests are generally used
        Measure and predict performance
        Most aptitude tests are standardized
            Must be reliable (consistent results)
                Inter-rater reliability
                Test-retest reliability
            Must be valid (measures what it is supposed to)
                Predictive validity
                Concurrent validity
    The IQ test is most widely used aptitude test
        Score based on mental and chronological age
        Scores may range from mental retardation to gifted

IQ may not represent changes that occur with age
Crystallized intelligence - increases with age
Fluid intelligence - decreases with age
It is argued that these tests are not sensitive to cultural groups
Intelligence and its relation to genetics and environment
Identical twins raised apart and fraternal twins are used to study this
African-Americans average 10 to 15 points lower on IQ tests than whites
Hereditary differences within groups
Gap is narrowing

Creativity: Cultural Differences, Assessment, and Individual Differences
The ability to think up new and useful ways to solve problems
Creativity involves divergent thinking
Tests are used to asses creativity
Divergent Production Test
Remote Association Test (RAT)
Creative people are less conforming to social norms and take more risks
External rewards make people less creative
Draw on highly organized knowledge in field of expertise

Multiple-Component Theories
Gardner's six separate intelligences
IQ tests do not test all of these intelligences
Linguistic
Logic-mathematical
Spatial
Musical
Bodily-kinesthetic
Personal
Gardner claims that the effects of brain damage support his theory
Sternberg's three aspects of intelligence
Componential aspect is covered by traditional IQ tests
Experiential aspect
Ability to combine experiences in insightful ways
Contextual aspect
Emphasizes adaptation to environment

## DISCUSSION QUESTIONS AND EXERCISES

**Note: These questions and exercises ARE the learning objectives for this chapter. Answer them accurately in your own words and you will have mastered the most important material. We guarantee it.**

1. Scope and History

a. The interdisciplinary effort of psychologists and others to understand how computers, animals, and humans acquire, represent, and use information have created what branch of science? How do they view computers, animals, and humans?

b. Explain briefly the causes of the downfall of behaviorists. What was the result of this downfall?

2. Language

a. Define **human language**. What are the three properties of human language? Give a brief description of each property.

b.    How unique are human language abilities?

3.    <u>Thought</u>

a.    What is the Sapir-Whorf hypothesis?  What research has been done to investigate this hypothesis?  What does the research show?

b.    What has research found concerning the forming, scanning and manipulation of images to solve problems?

c.    According to researchers, what two methods are used to learn **family-resemblance** categories?  When do we use these methods?

d.  Explain how a person would know that a robin is a bird according to the prototype view.

4.  Problem Solving

a.  Define **problem solving**. What are the three major steps in the problem solving process? Give a brief description of each step.

b.  What is **functional fixedness**? Give an original example.

c.      What is the difference between experts and novices when they are making plans and forming subgoals?

d.      Characterize each of the following methods for making decisions and evaluating strategies:

(1)     Trial and error

(2)     Hypothesis testing

(3)     Algorithms

(4)     Heuristics

5. Intelligence: Assessment, Controversies, and Theories

a. Identify and describe two different tests that are used to measure intelligence. What does standardized mean?

b. The requirements of any test are that it be reliable and valid? What does this mean and give an example?

c. Give two types of consistency a test must have. How do you determine the validity of a test?

d. Who published the first useful IQ test? Terman devised a formula that determined your intelligence quotient (IQ). Write down the formula below and put in some original numbers. Was this person mentally retarded or gifted?

e.	Several controversies have arisen over the use of IQ tests. Identify and describe at least two controversies concerning the use of IQ tests.

f.	Contrast **fluid intelligence** and **crystallized intelligence**.

g.	What procedures have been used to determine genetic relatedness in humans?

h.	In the US, African-Americans average 10 to 15 points lower on IQ tests than white people. Some researchers say this is because African-Americans are innately less intelligent. What arguments are given to rebut that conclusion?

6. <u>Creativity: Cultural Differences, Assessment, and Individual Differences</u>

a.   Define **creativity**. How is it measured, convergently or divergently? Identify and describe some of the popular tests that measure this ability.

b.   What are some characteristics of creative people. Do you possess any of these characteristics?

7. <u>Multiple-Component Theories</u>

a.   Gardner has proposed that there are six independent "intelligences". What are these independent "intelligences"? What supports his theory according to him?

b.      Sternberg advances a multiple-component theory that includes three aspects of intelligence. Identify and briefly describe each aspect.

## POSTTEST

1.      According to Sternberg's 3-part theory of intelligence, contextual aspects of intelligence refer to:
    a.      adaptation to the environment
    b.      analytic thinking that is covered by traditional IQ tests as well as planning and evaluation strategies
    c.      situational variables that affect how one attempts to solve a problem
    d.      combining experiences in insightful ways

2.      People who speak different languages think differently according to which of the following?
    a.      imaginal thought
    b.      linguistic resemblance hypothesis
    c.      propositional thought
    d.      Sapir-Whorf hypothesis

3.      Interested in learning how to be an sailor, Matt has just taken a test designed to predict how well he is likely to do in a sailor training program. Matt has taken a(n):
    a.      achievement test
    b.      aptitude test
    c.      intelligence test
    d.      test of general mental ability

4.      According to Terman's formula (which is based on Binet's concepts), if an 11-year-old child had an IQ score of 118, the child's chronological age would be:
    a.      11.5
    b.      12
    c.      13
    d.      15

5. An implication of the belief that intelligence is largely a function of genetics is:
   a. that an intellectually stimulating environment can increase a person's intellectual potential
   b. that educational opportunities should be made available to everyone
   c. that everyone has the potential to do well intellectually if given the chance
   d. that you either are intelligent or you are not and no environmental circumstance can change that fact

6. The type of thinking designed to overcome obstacles that block attainment of a goal is referred to as:
   a. functional fixedness
   b. imaginal thought
   c. problem solving
   d. propositional thought

7. By definition, language:
   a. involves a system of symbols that can be used to create an infinite variety of messages
   b. is used by many animal species
   c. refers to a system of vocal communications only
   d. stipulates symbols that mean the same thing to everyone

8. {T  F} Creativity requires very high intelligence.

9. At the end of her math course, Cindy takes a test to determine how well she has mastered the material.  Her math test is primarily:
   a. a test of her math potential
   b. an achievement test
   c. an aptitude test
   d. an intelligence test

10. Which of the following statements about concept formation is TRUE?
    a. concept formation has been demonstrated in both animals and humans
    b. concept formation is a basic grouping strategy used by animals and deficient humans
    c. concept formation is displayed mainly by intellectually gifted humans
    d. humans form concepts but other animals can not

11. Which of the following is NOT true of algorithms?
    a. they are step-by-step procedures for solving problems
    b. they are the quickest and most efficient way to solve any problem
    c. they can only be applied to some problems
    d. they consider all possible solutions to a problem

12. A heuristic is:
    a. a flash of insight
    b. a general problem-solving strategy
    c. a violation of the means/end strategy
    d. a way of making a compensatory decision

13. The main difference between concurrent validity and predictive validity is:
   a. the nature of the criterion measure
   b. the nature of the exam questions
   c. the number of subjects in which the validity is calculated
   d. the time period when the validity criterion is collected

14. "I before e except after c" illustrates which of the following?
   a. a morpheme
   b. a phoneme
   c. a phonological rule
   d. an orthographic rule

15. After seeing your new neighbor walking very stiffly by your house wearing horn-rimmed glasses on a chain, a cardigan sweater, and her hair in a bun, you decide she must be a librarian. Your judgment is based on:
   a. expected value
   b. subjective validity
   c. the availability heuristic
   d. the representative heuristic

16. As one looks into a crowded parking lot, the cars and trucks all look a little different, yet they all satisfy our concept of "personal vehicle." This notion that concept membership can be determined by similarity of types of features, as well as shared features, is demonstrated by the term:
   a. availability
   b. concept formation
   c. family resemblance
   d. mental set

17. Confirmation bias refers to the tendency to:
   a. choose the sure thing when a decision is framed in terms of probability of success
   b. seek supportive rather than contradictory information after making a decision
   c. seek supportive rather than contradictory information before making a decision
   d. take risks when the probability of success is high

18. Phonemes are the smallest units of _____ in a spoken language; morphemes are the smallest units of _____ in a language.
   a. meaning; sound
   b. meaning; syntax
   c. sound; meaning
   d. sound; syntax

19. Which, if either, of the following children would be considered the more intelligent: Greg is 6 years old and has a mental age of 8. Molly is 10 years old and has a mental age of 12.
   a. Alberta
   b. Patrick
   c. there is not enough information to make a judgment
   d. they would be considered equally intelligent because they each have a mental age two years higher than their chronological age

20. Watson and Crick built models of DNA, then Wilkins and Franklin tested those models against X-ray data. This problem solving method illustrates:
    a. algorithms
    b. heuristic analysis
    c. hypothesis testing
    d. trial and error

21. If a test accurately measures what it was designed to measure, we would say that the test is:
    a. consistent
    b. normative
    c. reliable
    d. valid

22. Intelligence is defined as:
    a. how much one has learned up to a particular time
    b. one's quantity of knowledge
    c. the capacity to learn and use information
    d. the quality that distinguishes humans from other animals

23. Mary found some old bricks in a field near her house. Some she used with some boards to make shelves; others she crumbled to make ground cover for her garden. Mary seems to have demonstrated:
    a. convergent thinking
    b. divergent thinking
    c. functional fixedness
    d. object imagery

24. The ability to make inferences, which is dependent on good neural functioning, is _____ intelligence.

25. Which of the following is NOT a dimension of expertise?
    a. enhanced memory due to organizing information into chunks
    b. increased ability to recognize analogies
    c. more effective use of planning
    d. superior intelligence

26. The man first responsible for developing the first intelligence tests designed to predict the school performance of children was:
    a. Alfred Binet
    b. David Wechsler
    c. Jean Piaget
    d. Lewis Terman

27. The abilities which are specific, such as one's vocabulary, are called _____ intelligence.

28. The field of study that concentrates on how knowledge is acquired, represented, and transmitted is called
    a. cognitive psychology
    b. cognitive science
    c. neuroscience
    d. neural psychology

29. Which of the following most accurately describes one of the goals of artificial intelligence research?
    a.    to have computers perform intelligent tasks
    b.    to have computers think like humans
    c.    to have humans think like computers
    d.    to teach computers to be intuitive and subjective

30. Problems that require a common object to be used in an unusual way may be difficult to solve because of:
    a.    functional fixedness
    b.    irrelevant information
    c.    mental set
    d.    unnecessary constraints

31. High reliability in a test means:
    a.    everyone got a high score on the test
    b.    people get about the same score they got before, when they retake it
    c.    the majority of examinees give the test a high rating for fairness
    d.    the test really does measure what it is supposed to measure

32. The smallest meaningful units of written language are _____.

33. If a test has good test-retest reliability:
    a.    it accurately measures what it says it measures
    b.    it can be used to predict future performance
    c.    the test yields similar scores if taken at two different times
    d.    there is strong correlation between items on the test

34. _____ is to manipulation of symbols (e.g. concepts, images, abstractions that stand for objects, events, etc.), as _____ is to a system of communicative gestures.
    a.    intelligence; language
    b.    language; intelligence
    c.    language; thought
    d.    thought; language

35. {T  F} Though not genetically identical like identical twins, fraternal twins are more genetically alike than other non-identical twin siblings.

36. Phonemes:
    a.    are the smallest units of sound in a spoken language
    b.    are the smallest units of meaning in a spoken language
    c.    are the rules of grammar in a spoken language
    d.    lie at the peak of the language hierarchy

37. Scientist who study animal language have concluded:
    a.    animals do not communicate with language
    b.    animals usually use language the way humans do
    c.    it is more complex than was originally thought
    d.    they share all properties of human languages

38. Intelligence tests measure _____ thinking; tests of creativity measure _____ thinking.
   a. convergent; convergent
   b. convergent; divergent
   c. divergent; convergent
   d. divergent; divergent

39. You can't think of a single instance in which Marcia helped you out, and so you decide that Marcia must be an ungenerous person. Your judgment is based on:
   a. expected value
   b. subjective validity
   c. the availability heuristic
   d. the representative heuristic

40. {T F} Human beings are the only species that communicate with each other.

41. A listener or reader who actively tries to understand the intended meaning that the speaker or writer is attempting to convey illustrates which property of human language?
   a. creative usage
   b. functionality
   c. interpersonal
   d. meaningful/referential

42. A person with an IQ of _____ would be classified as mentally gifted.
   a. 34
   b. 90
   c. 100
   d. 140

43. Which of the following traits is related to creativity?
   a. high dependence on others
   b. nonconformity
   c. very high intelligence
   d. working for extrinsic motivation

44. The main goal of the "culture fair" tests is to develop instruments that:
   a. eliminate the current differences between males and females on IQ tests
   b. lower the IQ scores of white males to approximate those of minorities
   c. use material people in different cultures have had an equal chance to learn
   d. yield the same average IQ score for all cultures

45. The fact that two people taking the same test in two different places will receive the same instructions, the same questions, and the same time limits means the test has been:
   a. regulated
   b. standardized
   c. synchronized
   d. validated

**POSTTEST ANSWERS**

1. a
2. d
3. b
4. c
5. d
6. c
7. a
8. F
9. b
10. a
11. b
12. b
13. d
14. d
15. d
16. c
17. b
18. c
19. a
20. c
21. d
22. c
23. b
24. fluid
25. d
26. a
27. crystallized
28. b
29. a
30. a
31. b
32. letters
33. c
34. d
35. F
36. b
37. c
38. b
39. c
40. F
41. c
42. d
43. b
44. c
45. b

# PSYCH JOURNAL

**Please use the following pages to record your thoughts and feelings about the following questions.**

1.   Does this chapter help you understand the ways in which you have solved problems in the past? Does it suggest new ways that you might approach problems in the future? Think how James Watson solved the mystery of DNA. He drew on what Sternberg calls contextual intelligence to put himself in the right place at the right time. He used visual images as well as words in his thought processes. He systematically tested hypothesis longer than he should have; he let them go when evidence contradicted them. Have you used any of these methods?

_____

_____

_____

_____

_____

_____

_____

_____

_____

_____

_____

_____

_____

_____

_____

_____

_____

_____

2. Do you think that you SAT or ACT scores are an accurate predictor of how successful you will be in life? What other talents or strong points do you wish people knew about you before they predicted your chances of success?

_____

_____

_____

_____

_____

_____

_____

_____

_____

_____

_____

_____

_____

_____

_____

_____

_____

_____

_____

3. Has anyone ever offended you by using what you consider to be an offensive term to refer to a group of individuals to which you belong, such as calling a professional woman "honey"? Do you think "political correctness" is a positive movement in our culture, or people just being too sensitive?

_____

_____

_____

_____

_____

_____

_____

_____

_____

_____

_____

_____

_____

_____

_____

_____

_____

_____

_____

_____

_____

4. Do you think computers will ever by able to think like humans enough to replace human beings in the workplace. Do you think this would be a positive development?

_____

_____

_____

_____

_____

_____

_____

_____

_____

_____

_____

_____

_____

_____

_____

_____

_____

_____

_____

_____

# Chapter 8 -- Infancy and Childhood

## CHAPTER SUMMARY

1.      Four major themes are important in the study of human development: Most modern psychologists recognize that a continuous interaction between heredity and environment determine important aspects of development. Continuity and discontinuity describe the process of slow growth over time and the developmental surges that are made in transition from one stage of development to another. Early experience and late experience focus on the character of development at different ages. Unidirectional pressures and reciprocal pressures draw attention to how the child affects the agents who influence development.

Five approaches to psychology are functionalism (combines the study of children with the application of those studies); the psychodynamic approach (emphasizes the interaction of heredity and environment); behaviorism (focuses exclusively on the affects of environment on observable behavior); the ethological/organismic approach (focuses on inherited biological predispositions that change systematically during the course of development); and the lifespan approach (focuses on continuity and change throughout life). Two methods dominate developmental research. Longitudinal research involves periodically testing the same person or group of people over a significant period of time. In cross-sectional research investigators compare the responses of people of different ages, education, ethnic background and/or economic status; all are studied at the same time.

2.      Prenatal development refers approximately to the 9-month period of development before birth. The first 2-week period of development after fertilization is called the period of the ovum. A fertilized ovum divides and plants itself in the wall of the uterus. The period of the embryo from the third week through the eighth week is characterized by extremely rapid growth. From the third month through birth the organism is called a fetus. Average birth weight is about 7 pounds, and average length is about 20 inches.

3.      An infant grows about 1 inch per month in the first six months. Infants are born with an incomplete nervous system that develops rapidly after birth. Development of myelin sheaths helps them gain control over body parts. Infants are born with rooting and sucking reflexes that are essential for getting milk from a breast or bottle. Motor development is marked by the disappearance of certain reflexes, and the development of voluntary movement, such as sitting up, crawling, and standing. Culture and experience influence when infants develop certain voluntary movement. Experiments indicate that infants perceive depth as early as six months and size as soon as nine months after birth. They also prefer patterned to unpatterned stimuli.

4.      Language is most likely not the result of imitation or operant conditioning. According to Chomsky, language acquisition is probably due to children learning cognitive representations that include units of thought and rules for combining those units. Children may learn languages more easily during a critical period between the ages of 2 and 13. Childhood is most notably marked by increases in cognitive development. Piaget studied two processes in children that account for much of this development: assimilation and accommodation. As a result of his years of observation

of children, he divided cognitive development into a series of sequential stages: sensorimotor (birth to 2 years); preoperational (2 to 7 years), concrete operational (7 to 11 years), and formal operational (about 13 years and older).

Children as young as three years old understand that thinking is not the same as seeing, talking, or acting.  An important part of personality and social development in infancy is attachment.  Healthy and secure attachments to parents or caregivers have been associated with higher levels of cognitive development.  Through imitation and reinforcement children acquire personality and social behaviors.  Parents help transmit cultural standards for gender roles both by treating boys and girls differently from an early age, and by modeling different behaviors for each sex.  Teachers, peers, and siblings influence social development.

5.      Children who seem to live in a world of their own, avoid social interactions, have language delays, and develop rigid routines suffer from infantile autism.  Treatment for autism is still in the experimental stage.  The term schizophrenia is applied to a range of disorders that are characterized by disorganization of thoughts, perceptions, communication, emotions and motor activity.  Schizophrenic children are more socially responsive and emotionally dependent than autistic children.  Treatment varies and the chances of being cured are poor.  An attention deficit disorder is indicated when children seriously lag behind others in the ability to concentrate on a task or activity.  Children with this disorder are often treated with stimulant drugs and cognitive-behavioral therapy.  Culture is a factor in diagnosing childhood disorders.

## KEY TERMS AND CONCEPTS

The History and Scope of the Study of Human Development
    Human development
        Genes are inherited from our parents, and expressed as characteristics
    Goals of developmental psychology
        Describe and explain this process in terms of general principles
    Four major themes in the study of human development
        Heredity and environment
        Continuity and discontinuity
        Early experience and late experience
        Unidirectional pressures and reciprocal pressures
    Five approaches to psychology
        Functionalism
            Combination of study of children and application of studies
        Psychodynamic approach - heredity and environment interaction
            Suffered from weak research methodology
        Behaviorism - affects of environment on observable behavior
        Ethological-organismic approach
            Inherited predispositions that change during development
        Life-span approach - continuity and change throughout life
    Two methods of developmental research
        Longitudinal research
            Periodic testing over long span of time
            Advantages and disadvantages
        Cross-sectional research
            Comparison of people at different ages
            Advantages and disadvantages

Prenatal Development
 The 9-month period of development before birth
  First 2-week period is period of ovum
   Fertilized ovum divides and plants itself in wall of uterus
  Third week through eighth week is period of embryo
   Characterized by extreme rapid growth
   Embryo is dependent on the placenta
  Third month through birth the organism is called a fetus
   Average birth weight is about 7 pounds
   Average length is about 20 inches

Birth to 1 Year
 First six months
  Infant grows about 1 inch per month
  A gain of about 5 to 7 ounces a week occurs
  Abnormal gains in these areas warn of health problems
 Infants are born with an incomplete nervous system
  Develops rapidly after birth, slows down during 2nd and 4th year
  Myelin sheaths help gain control over body parts
 Rooting and sucking reflexes are present at birth
  Essential for getting milk from a breast or bottle
  Four other common reflexes
   Grasp reflex - close fingers around object
   Dancing reflex - prancing with legs
   Moro reflex (startle) - reaction of legs with loss of support
   Tonic neck reflex - cause movement of arms when on their back
  Lack of these reflexes indicates abnormalities in nervous system
  Culture and experience influence certain voluntary movements
 Depth perception occurs as early as six months
 Size perception occurs as early as nine months
  Patterned is preferred to unpatterned stimuli

Childhood
 Growth slows down until end of childhood, when it increases again
 Theories of language development
  Social learning theory - imitation of adults
  Operant conditioning - reinforcement plays a role
  Chomsky - language acquistion due to learning cognitive representations
   Include units of thought and rules for combining those units
  Critical period for learning languages
   Between the ages of 2 and 13
 This period is most notably marked by increases in cognitive development
  Piaget's accommodation and assimilation
  Piaget's series of sequential stages:
   Sensorimotor - birth to 2 years
    Object permanence - objects continue to exist
   Preoperational - 2 to 7 years
   Concrete operational - 7 to 11 years
   Formal operational - about 13 years and older
 Children young as three, know difference between thinking, seeing, and acting
 Heredity and environment influence smiling
 Attachment is important part of personality and social development
  Imprinting - attachment to first moving object seen
  Healthy and secure attachments associated with higher cognition

Personality and social behaviors acquired through imitation and reinforcement
    Identification - assuming other's characteristics
Gender roles play a role in cultural standard transmission
    Teachers, peers, and siblings also influence social development

Childhood Disorders
    Infantile autism - avoidance of social interactions, language delays
        Treatment is still in experimental stages
    Schizophrenia - disorganized thoughts, perceptions, and motor activity
        More socially responsive and emotionally dependent than autistic child
        Treatment varies, chances of being cured are poor
    Gender identity disorder - child adopts role and outlook of opposite sex
        More often found in males
    Attention deficit disorder - lagging behind others in abilities
        Treatment is with stimulant drugs and therapy
    Culture plays a role in diagnosing childhood disorders


## DISCUSSION QUESTIONS AND EXERCISES

**Note: These questions and exercises ARE the learning objectives for this chapter. Answer them accurately in your own words and you will have mastered the most important material. We guarantee it.**

1.    <u>History</u> <u>and</u> <u>Scope</u> of <u>the</u> <u>Study</u> of <u>Human</u> <u>Development</u>

a.    Define **human development**. What are human development's goals?

b.    Characterize each of the following four themes that have been important in the study of human development:

    (1)    Heredity and environment

    (2)    Continuity or discontinuity

    (3)    Early and later experience

(4)    Unidirectional pressures and reciprocal interactions

c.    Characterize each of the following theoretical approaches to human development:

(1)    Functionalism

(2)    Psychodynamic approach

(3)    Behaviorism

(4)    Ethological/organismic approach

(5)    Life-span approach

d.    Contrast both **longitudinal research** and **cross-sectional research**. What are some of the advantages and disadvantages to both? Is there a single design for a single purpose? Why or why not?

2.	Prenatal Development

a.	What does **prenatal** refer to?  What happens during this period of development?
	Give a brief description of each of the three periods that occur.

b.	Identify some of the major developments that occur during the period of the
	fetus.

3.	Birth to 1 Year

a.	What are some of the norms associated with weight gain and growth during the
	first six months of life?  What do deficiencies in these areas indicate?

b.	In what ways is the nervous system at birth "unfinished"?

c. Identify some of the normal patterns associated with motor development. What does the presence or absence of reflexes indicate?

d. Identify and define three of the major reflexes given to you in your textbook. What does a lack of these at birth indicate?

e. A major accomplishment in the first year of life is the replacement of motor reflexes with voluntary motor control. What does this achievement mean? How does culture play a role in the development of voluntary motor control?

f. What is the device that Walk and Gibson designed to study infant depth perception? Describe the device and how it was used in the study. What did the study reveal?

g.     What do the looking preferences found in research tell us about babies' preferences?

4.     <u>Childhood</u>

a.     Trace the development of language, identifying some of the major landmarks that occur.

b.     There are three theories of language development presented in this chapter. Briefly describe each theories attributes and describe which one you might agree with.  Why or why not?

c.     What role does social deprivation play in language development?  What is the critical period?

d.      What is **schema**?  What role do schemas play in Piaget's theory?

e.      Define and give an example of **assimilation** and **accommodation**.

f.      Briefly characterize the four stages of cognitive development proposed by Piaget.  State the age range for each stage and identify at least three major characteristics.

(1)     Sensorimotor

(2)     Preoperational

(3)     Concrete operations

(4)     Formal operations

g. Both 3- and 4-year-old children know that a person can think about something without seeing it, talking about it, or doing anything with it.  What research has been performed to prove this?

h. What role does **attachment** play in personality and social development?  What are the series of three steps that human attachments normally develop during the first 6 months of life?

i. Describe the strange situation procedure.  Identify and briefly explain the two categories of attachment.

j. Define **identification**.  What term do modern psychologists use to describe the copying of others behavior?  Identify another important mechanism for shaping personality and social development.

k.      Mothers and fathers influence almost every aspect of their children's personality and social development.  What is one area of parental influence described in your textbook?  Why is this area important?

l.      How do teachers, siblings, and culture affect personality and social behavior?

5.    Childhood Disorders

a.      Identify some of the symptoms of **infantile autism**.  Give a brief description of some of the possible causes, and possible treatments.

b.    Characterize **schizophrenia**.  What is the difference between schizophrenic children and autisic children?  What treatments are available?

c.    What is **gender identity disorder**?  How does this disorder happen?  What treatments are available?

d.    Identify some of the main characteristics of **attention deficit disorder**.  What are its causes and cures?

e.    What role does culture have in diagnosing childhood disorders?  What problems can arise?

## POSTTEST

1.    Ronnie's spouse has just gotten a big promotion at her job.  Ronnie is as proud and excited as if he had gotten the promotion himself.  This illustrates which of the following?
      a.    identification
      b.    modeling
      c.    reaction formation
      d.    reinforcement

2.    A child refuses to crawl out over a glass platform that extends over a several foot drop-off.  This test demonstrates that the child:
      a.    is able to control visual accommodation
      b.    is able to crawl
      c.    is able to perceive depth
      d.    has had visual cliff training

3.    Which model of development emphasizes the role of heredity in one's development?
      a.    behaviorism
      b.    empiricism
      c.    humanism
      d.    nativism

4.    Faced with having to learn how to work a combination lock for his bicycle and having had no previous experience with such locks, Billy must use the thinking process that Piaget called:
      a.    accommodation
      b.    assimilation
      c.    cognitive analysis
      d.    operationalization

5. To successfully read a college level business text, a student should have reached at least the _____ stage of intellectual development.
   a. concrete operations
   b. formal operations
   c. preoperational
   d. sensorimotor

6. Brian and Ken are both interested in the development of a sense of humor. Brian tested the sense of humor in a group of six-year-olds and then retested the same children every four years until they were twenty-two years old to see how their sense of humor changed. Ken tested eight different groups of children -- ages 3, 5, 8, 9, 11, 14, 15, and 17-- to see if there were any differences in their sense of humor. Brian is doing _____ research; Ken is doing _____ research.
   a. cross-sectional; experimental
   b. cross-sectional; longitudinal
   c. experimental; cross-sectional
   d. longitudinal; cross-sectional

7. Professor Chomsky believes humans are born with a "language acquistion device" that allows their speech to exhibit proper syntax rules as they get older. He probably is associated with which developmental position?
   a. behaviorism
   b. empiricism
   c. functionalism
   d. nativism

8. {T  F} The process by which a young fowl attaches itself to the first object it sees is called attachment-in-the-making.

9. A _____ _____ is society's approved ways for men and women to behave.

10. A mental structure that organizes responses to experience is called a _____ .

11. An infant will turn his head toward anything which touches his cheek. This demonstrates the _____ reflex.
    a. grasping
    b. Moro
    c. rooting
    d. dancing

12. A child who knows how to ride a bicycle would probably find it relatively easy to learn how to ride a motorcycle. Piaget would call this thinking process:
    a. accommodation
    b. assimilation
    c. cognitive analysis
    d. maturation

13. A sudden loud noise made in the vicinity of the newborn child will give rise to the _____ reflex.
    a. Babinski
    b. headturning
    c. Moro
    d. sucking

14. A pure environmental approach to development would be characterized as one of:
    a. discontinuity
    b. nativism
    c. reciprocal interactions
    d. unidirectional pressures

15. Which of the following is an advantage associated with the cross-sectional method of analysis in the study of development?
    a. it is time efficient
    b. it provides good evidence of individual changes in behavior over time
    c. it requires only one or two subjects
    d. it is inexpensive

16. When do infants usually form attachments to their parents?
    a. between age one and one-and-a-half
    b. between six and twelve months of age
    c. by the end of the first month of life
    d. during the first six months of life

17. Which of the following supports the case that social smiling is genetically based?
    a. parents who do not smile often have babies who seldom smile socially
    b. most identical twins show social smiling at the same age
    c. smiling has critical periods of development
    d. the smiles are originally reflexes

18. Self injurious behavior is associated with which of the following disorders?
    a. attention deficit disorder
    b. childhood schizophrenia
    c. gender identity
    d. infantile autism

19. When children wonder why the ocean never stops to rest, they are demonstrating a belief that all things are living, a belief called:
    a. animism
    b. centration
    c. existentialism
    d. pluralism

20. Which approach to the study of development views one's early experiences as a more critical period than later (e.g., adult) experiences?
    a. behaviorism
    b. functionalism
    c. life span
    d. psychodynamic

21. Which of the following has Piaget's stages of cognitive development in the correct chronological order?
    a. sensorimotor; preoperational; concrete operations; formal operations
    b. preoperational; sensorimotor; concrete operations; formal operations
    c. preoperational; formal operations; sensorimotor; concrete operations
    d. sensorimotor; concrete operations; formal operations; preoperational

22. According to Piagetian theory; plans, strategies, and rules for solving problems and classifying are called:
    a. concepts
    b. mental sets
    c. operations
    d. schemata

23. Most motor behaviors come under voluntary control _____.
    a. by the end of the 6th month
    b. by the end of the first month after birth
    c. by the end of the second year
    d. during the 1st year of life

24. According to Piaget, during which stage of cognitive development do children come to realize that an object continues to exist when they cannot see it or touch it?
    a. concrete operations
    b. formal operations
    c. preoperational
    d. sensorimotor

25. Internal organs begin to develop during which prenatal period?
    a. embryonic
    b. fetal
    c. ovum
    d. zygotic

26. Most children younger than eight months of age do not "understand" or enjoy "peek-a-boo" because they have not developed:
    a. abstract thinking
    b. conservation
    c. object permanence
    d. reversible thinking

27. Arnold Gesell used _____ methods to establish _____ development.
    a. behavioral; environmental influences on
    b. behavioral; stage of
    c. ethological; environmental influences on
    d. ethological; stages of

28. A disorder that can occur in childhood, when a child rigidly adopts the role and outlook of the opposite sex is _____ _____.

29. The period of the ovum refers to the first _____ after birth.
    a.  month
    b.  two-weeks
    c.  two-months
    d.  two years

30. Because of its underdeveloped brain, a human infant would likely benefit MOST from what type of environment during the first two years of life?
    a.  a normal, non-enriched environment to allow for brain development
    b.  a stimulating, enriched environment to increase interconnections and cortical development
    c.  a subdued environment from a stimulation perspective
    d.  an environment that doesn't tax the baby's abilities

31. The assumption of a preoperational child that the reason a car is moving is because she is in it is an example of:
    a.  centration
    b.  conservation
    c.  egocentric thinking
    d.  reversible thinking

32. The longest stage of prenatal development is the:
    a.  embryonic stage
    b.  fetal stage
    c.  germinal stage
    d.  zygote stage

33. Which of the following is a weakness associated with longitudinal research?
    a.  cultural experiences cannot be controlled
    b.  it cannot measure the stability of behavior
    c.  it costs a lot in terms of time and money
    d.  historical influence cannot be controlled

34. An appealing toy is placed 4 feet away from a month-old infant but the infant does not try to get it because:
    a.  infants are not interested in toys
    b.  the infant cannot see it clearly
    c.  the infant doesn't know what to do with it
    d.  the infant has insufficient motor skills to get to it

35. The strange situation procedure, in which researchers unobtrusively watch an infant in the presence or absence of several combinations of the child, caretaker, and stranger, is used to study:
    a.  attachment
    b.  dependence and independence
    c.  identification
    d.  social smiling

36. How soon after birth can infants perceive distance or depth?
    a.  between one and two weeks
    b.  by the end of the first day
    c.  by the sixth week
    d.  by the time they begin to crawl

37. Greg is presented with 4 dogs and 2 cats and is asked whether there are more dogs or more animals. If he answers correctly that there are more animals, it is likely that he is in the:
    a. concrete operations period
    b. preoperational period
    c. formal operational stage
    d. sensorimotor period

38. Researchers have discovered that during specific periods in one's life, certain abilities must develop or they will not develop later. These are known as:
    a. critical periods
    b. developmental milestones
    c. imprinting
    d. maturational readiness

39. The newborn engages in a variety of behaviors (such as head turning, sucking, and sneezing) in response to specific stimuli. These behaviors are called _____ responses.
    a. automatic
    b. conditioned
    c. reflexive
    d. unintentional

40. _____ is most famous for this theory that all children go through a series of sequential intellectual stages.
    a. Ainsworth
    b. Bowlby
    c. Gesell
    d. Piaget

41. Which type of drugs are most commonly used to treat attention deficit disorder?
    a. antidepressants
    b. depressants
    c. stimulants
    d. tranquilizers

42. {T F} The filter used to exchange food and wastes between the embryo and mother is called the placenta.

43. Which of the following developmental sequences is in correct order?
    a. embryo, zygote, fetus, infant
    b. fetus, embryo, ovum, infant
    c. ovum, embryo, fetus, infant
    d. ovum, fetus, embryo, infant

44. The ability to perform mental operations separates Piaget's _____ stage from his _____ stage.
    a. concrete operations; formal operations
    b. preoperational; concrete operations
    c. sensorimotor; concrete operations
    d. sensorimotor; preoperational

45. Mr. Smith takes care of his and Mrs. Smith's children while she supports the family. As a caretaker, Mr. Smith takes excellent care of the kids' physical and stimulation needs. He has activities scheduled for the entire day; so many that the kids frequently are in bed exhausted before Mrs. Smith returns from work. Ainsworth would predict which type of attachment between the Smith's kids and their dad?
    a.    apathetic relationship
    b.    anxiously attached
    c.    securely attached
    d.    strong dependent relationship

## POSTTEST ANSWERS

1.  a
2.  c
3.  d
4.  a
5.  b
6.  d
7.  d
8.  F
9.  gender role
10. schema
11. c
12. b
13. c
14. d
15. a
16. d
17. b
18. d
19. a
20. d
21. a
22. c
23. d
24. d
25. a
26. c
27. d
28. gender identity disorder
29. b
30. b
31. c
32. b
33. c
34. d
35. a
36. a
37. a
38. a
39. c
40. d
41. c
42. T
43. c
44. b
45. c

**PSYCH JOURNAL**

     **Please use the following pages to record your thoughts and feelings about the following questions.**

1.     Do you have any "natural" talents?  What skills and talents did your parents help you to develop?  Do you remember any particular toys, games, or books that helped you develop particular skills?

_____

_____

_____

_____

_____

_____

_____

_____

_____

_____

_____

_____

_____

_____

_____

_____

_____

_____

2. How did your early childhood experiences shape your character? Which grade school experiences do you think had an impact on you?

_____

_____

_____

_____

_____

_____

_____

_____

_____

_____

_____

_____

_____

_____

_____

_____

_____

_____

_____

3.  Ask your parents or consult your baby book to find out when you reached various developmental milestones such as smiling, talking, walking, etc. Was your development gradual, or did you go through bursts of intellectual, physical, athletic, or emotional growth?

_____

_____

_____

_____

_____

_____

_____

_____

_____

_____

_____

_____

_____

_____

_____

_____

_____

_____

_____

4.　Did you ever know a child with autism, schizophrenia, gender identity disorder, or hyperactivity? What were your impressions of him or her at the time? Do these terms help you understand a child you knew when you were young who you thought was just "weird"?

_____

_____

_____

_____

_____

_____

_____

_____

_____

_____

_____

_____

_____

_____

_____

_____

_____

_____

_____

_____

# Chapter 9 -- Adolescence, Adulthood, and Aging

## CHAPTER SUMMARY

1.      The life-span developmental approach is concerned with the description and explanation of changes in behavior within an individual and differences between individuals from conception to death.

2.      The physical changes include a growth spurt, puberty, and the development of primary and secondary sex characteristics. Height and weight are largely determined by heredity. The first viable sign of puberty for girls is menarche and for boys, growth of the scrotum and the appearance of pubic hair. Kohlberg identified six stages of moral reasoning. It is not clear whether these stages are valid for non-Western cultures or for women. According to Erikson, the identity crisis is an important stage in the personality development of adolescents. The timing of physical maturation and cultural context affects our sense of ourselves. Peers have a powerful influence on adolescents in part because they listen as well as talk to each other. Parents and others tend to try to impose their own views. Sexual activity among U.S. teenagers is high. Even the occurrence of AIDS has not greatly reduced teenage sex.

3.      One way to look at young adulthood is to analyze the commitments that are made at this time: moral commitments (Kohlberg's postconventional stage is usually reached at this point); interpersonal commitments (Erikson's stage of intimacy versus isolation occurs in which marriage and family commitments are a major theme); and mastery commitments (the establishment of and mastery in a career). Levinson's framework for development in young adulthood contains several stages, including a transition stage at about age 30, and a settling down stage, at about 40 years of age.

4.      Middle age is concerned first with Erikson's generativity versus stagnation stage. In the same way, Levinson's framework at this age includes a kind of midlife transition. Physiological capacities such as muscle strength, lung capacity, cardiac output and other functions decline. Most middle-aged people have about 20 years of marriage behind them at this age, and they find themselves sandwiched between the needs of two generations: their children and their aging parents. In regard to occupational commitments, people tend to be either persisters (staying with one career throughout life) or shifters (changing careers in the middle years).

5.      Late adulthood involves many physical changes which affect sensory abilities as well as general health and mobility. Aging is determined by heredity factors, personal factors, and health factors. Cognitive changes occur in recalling recently learned material, manipulating mental symbols, and dividing attention. Family commitments include being a grandparent. Most elderly couples enjoy an increase in marital satisfaction. Retirement is a major aspect of late adulthood. Few studies have been done on minorities and retirement. Some conclusions, such as retirement for health reasons, generalize across ethnic groups; however, others do not. For example, middle-class whites define themselves as retired when they no longer hold a full-time job. Many African Americans and Mexican Americans do not have full-time jobs; they "retire" when they are too disabled to do the kind of work they have done in the past.

6.      Psychologists have outlined distinct phases of grieving which include shock and overwhelming sorrow, coping with anxiety and fear, obsessional review of how the death might have been prevented, and recovery. Kübler-Ross has outlined five

stages in the adjustment process to one's own death:  denial and isolation, anger, bargaining, depression, and, finally, acceptance.  Kübler-Ross's stages are based on her observations; everyone does not go through these stages and patients should not be manipulated to follow them.  Society is beginning to acknowledge the individual's wishes in the matter of death.  Euthanasia has become a controversial issue.  Living wills, that allow patients to make decisions about withholding or withdrawing medical treatments, are recognized in some states.


## KEY TERMS AND CONCEPTS

History and Scope of Life-Span Approach to Development
      Concerned with changes in behavior from conception to death

Adolescence
      Extends from about age 12 to the late teens
      Passage from childhood to adulthood
      Physical changes
            Growth spurt--growth rate for girls increases earlier than boys
            Puberty
                  Menarche is first sign for girls
                  Growth of scrotum and pubic hair is first sign for boys
            Primary and secondary sex characteristics
            Height and weight determined by heredity
      Cognitive and moral development
            Principles of right and wrong
            Teenagers require less supervision because of internalization
            Kohlberg's six stages of moral reasoning (Western cultures)
      Personality development
            Erik Erikson's personality stages - eight stages of life
                  First four occur before puberty and last four after puberty
                  Identity crisis - occurs during puberty
            Timing of physical maturation affects our sense of ourselves
      Social Changes
            Peers begin to have an influence on adolescents
            Parents try to impose own views
      Intimacy and Sexuality
            Sexual activity is high
            AIDS has not affected rate of sexual activity

Young Adulthood
      Moral, interpersonal, and mastery commitments
      Family commitments
            95 percent of U.S. citizens get married, trend toward fewer children
            Cohabitation - unmarried people living together
            Family cycle
      Occupational commitments
            White's four aspects of personality and social development
                  Respond to people warmly and respectfully
                  Base decisions upon their own beliefs
                  Respect cultural values
                  Care about society and the people in it
            Age 30 transition and becoming one's own man or women

Middle Adulthood
    Sequence and development in middle years
        Erikson's generativity versus stagnation stage
            Generativity - producing and contributing to world
            Stagnation - fulfilling expectations
        Levinson's midlife transition
    Physical changes
        Muscle strength, lung capacity, cardiac output decline
    Family commitments
        20 years of marriage behind them
        Needs of children and aging parents are experienced
    Occupational commitments
        Persisters or shifters

Late Adulthood
    Dramatic increase in elderly population
    Physical and mental abilities
        Many physical changes are affected along with general health
        Aging determined by heredity factors, personal factors, health factors
        Cognitive changes occur: dividing attention, recalling material
    Family commitments
        Being a grandparent
        Increase in marital satisfaction
    Occupational commitments
        Retirement becomes a major aspect
        Studies on minorities and retirement
            Middle-class whites - retire when no longer hold full-time job
            African/Mexican Americans - retire when too disabled to do job
        Adjustment to retirement
            Remain active to sustain physical health and cognitive skills
                Traveling and staying in touch with younger generation

Death and Dying
    Adjusting to the death of loved ones
        Initial response - shock and sorrow
        Coping with anxiety and fear
        Obsessional review of how death could have been prevented
        Recovery
    Accepting one's own death
        Kübler-Ross's five stages in adjustment process
            Denial and isolation
            Anger
            Bargaining
            Depression
            Acceptance
                These stages are not invariant or universal
    Euthanasia and living wills
        Euthanasia is a controversial issue
        Living wills - allow patients to make decisions

## DISCUSSION QUESTIONS AND EXERCISES

**Note: These questions and exercises ARE the learning objectives for this chapter. Answer them accurately in your own words and you will have mastered the most important material. We guarantee it.**

1. <u>History</u> <u>and</u> <u>Scope</u> of <u>Life-Span</u> <u>Approach</u> <u>to</u> <u>Development</u>

a. What is the **life-span developmental approach** concerned with?

2. <u>Adolescence</u>

a. Define **adolescence**. Describe some of the physical changes that occur during this period, like height and sexual characteristics. What is the growth spurt?

b. What is the first sign of puberty for both girls and boys?

c. Why do teenagers require less and less supervision as they grow older? What were some of the results of Piaget and Kohlberg's research on moral reasoning?

d.    What are some of the controversies regarding moral reasoning, cultures, and gender?

e.    Does personality develop throughout the life span?  Does it ever stop developing?

f.    Give a brief description of Erik Erikson's various stages of personality development.  What is the important stage in Erikson's theory?  Why?  What influences this stage?

g.    What has research revealed about the timing of physical maturation across cultures?

h.    What role do peers play during adolescence?

i.    Characterize the intimate relationship.  What does Erikson think about
      intimacy?  Has the rate of sexual activity among teenagers gone up or down?
      Does the threat of AIDS have anything to do with this?

3.    Young Adulthood

a.    Give a brief description of moral, interpersonal, and mastery commitments.
      What stages do each of them represent?

b.    What are some of the important statistics about marriage and having children
      during young adulthood?  What is the family cycle?  What are the major events?

204

c.     What are some of the sources of variability in patterns of family cycles?

d.     What three periods do men go through during young adulthood, according to Levinson?  Describe each period.  What do women go through during young adulthood?

4.     <u>Middle Adulthood</u>

a.     According to Erikson, what are the two feelings that can dominate during middle adulthood?  Characterize each one and how they can occur.

b.    Levinson proposed a universal sequence of periods during middle adulthood. Identify some of the major milestones he proposed.  Is there really a midlife crisis?

c.    What are some of the physical changes that occur during middle adulthood? What is the effect of these changes?  What can a middle-aged person do to help alleviate some of these changes?

d.    Describe family commitments in middle adulthood.

e.     Contrast shifters and persisters. Which category are your parents in?

5.     <u>Late</u> <u>Adulthood</u>

a.     What are the trends in the number of senior citizens in this country and others? How do these trends affect you?

b.     Give a brief description of the biology of aging. How does aging occur? What factors will determine your well-being as you age?

c.     Identify some of the cognitive and sensory changes that occur in senior citizens.

d. How do grandchildren affect grandparents and how do grandparents affect grandchildren? Does the gender of the grandchild affect the relationship?

e. When does marital satisfaction reach its highest level? Why? What is the down side to elderly husbands and wives?

f. Describe some of the differences between African Americans, Mexican Americans, and middle-class whites with regard to retirement.

6. <u>Death</u> <u>and</u> <u>Dying</u>

a. Psychologists are only beginning to learn about the grief process when adjusting to the death of a loved one. What are some of the phases of grieving given to you in your textbook? How can children, clergy, and caregivers help?

b. Elizabeth Kübler-Ross interviewed 200 people about how they felt about their own death. What are the five stages she identified in the adjustment process? Characterize and give an example of each stage.

c. Define **euthanasia**. Why is this so controversial? What is a living will? Do you have a living will?

**POSTTEST**

1.      {T  F} The age at which an individual reaches puberty can affect his/her
        personality and social development.

2.      Which of the following is a secondary sex characteristic?
        a.      menstruation in the female and ability to produce sperm in the male
        b.      the breasts of the female and large chest muscles of the male
        c.      the ovaries of the female and the testes of the male
        d.      the uterus in the female and the penis of the male

3.      A person who believes that she did not accomplish anything worthwhile with
        her life would likely be in which of Erikson's stages?
        a.      autonomy vs. doubt
        b.      ego-integrity vs. despair
        c.      generativity vs. stagnation
        d.      identity vs. role confusion

4.      Do most adolescents prefer to discuss problems with peers or with parents?
        a.      with both peers and parents, but on separate occasions
        b.      with parents only
        c.      with peers and parents simultaneously
        d.      with peers only

5.      The pattern of events that is repeated in each new generation is referred to as
        the:
        a.      developmental cycle
        b.      family circle
        c.      family cycle
        d.      family pattern

6.      The learning of inhibitions and self-control is called:
        a.      conscience development
        b.      ethics
        c.      internalization
        d.      morality

7.      Erikson posited with stages of development that occurred through the life span.
        Each of these stages was characterized by:
        a.      a nonevent transition
        b.      a psychological crisis
        c.      an anticipated life transition
        d.      an initiation rite

8.      The beginning of menstruation which marks the onset of puberty in girls is
        called _____.

9.  How did Kübler-Ross come to the conclusion that dying patients go through five stages of dealing with the knowledge that they are terminally ill?
    a.  by interviewing nurses who worked in the terminally ill wards of nearly a dozen hospitals
    b.  by interviewing over 200 terminally ill patients who knew they were going to die
    c.  by reading "death-bed" autobiographies
    d.  through her own personal experience

10. {T  F} Conventional wisdom once said that fear of AIDS would greatly suppress teen sex.  Research has shown that it is working.

11. Kohlberg found that the typical responses of adolescents to the moral dilemmas he presented them were at the _____ level.
    a.  amoral
    b.  conventional
    c.  preconventional
    d.  postconventional

12. Which one of the following is NOT one of the determinants of aging?
    a.  health factors
    b.  heredity factors
    c.  personal factors
    d.  psychological factors

13. What is a good predictor of marital success?
    a.  having an older sibling who is happily married
    b.  knowing each other since grade school
    c.  marrying during adolescence
    d.  quality of communication during dating

14. Henry, a 14-year-old, is suspicious of both friends and strangers, and very cautious before trying anything new.  According to Erikson, these traits were likely a result of Henry's inability to meet the crisis of:
    a.  autonomy vs. doubt
    b.  initiative vs. guilt
    c.  introversion vs. extroversion
    d.  trust vs. mistrust

15. Young people tend to think about death _____ older people.
    a.  as often as
    b.  less fearfully than
    c.  less than
    d.  more than

16. In our society, adolescence is generally considered to be from _____ to _____.
    a.  age 12; age 16 or 17
    b.  age 12; age 20 or 21
    c.  puberty; age 16 or 17
    d.  puberty; age 20 or 21

17.    The "initial response" to the death of a spouse is a phase characterized by:
       a.    an obsessional review of how the death could have been averted
       b.    review of old memories of the times experienced with the dead spouse
       c.    shock and overwhelming sorrow
       d.    worry of a nervous breakdown

18.    A middle-aged person who fails to successfully resolve his/her midlife crisis
       may "stagnate" according to Erikson.  Which of the following would NOT be
       true about a person who is "stagnant"?
       a.    she/he can not reverse the condition and become generative
       b.    she/he focuses on personal matters
       c.    she/he is not concerned about others
       d.    she/he is self-absorbed

19.    Which of the following is NOT one of the commitments that Lowenthal
       proposed as a framework for investigating development during young
       adulthood?
       a.    interpersonal
       b.    mastery
       c.    moral
       d.    relationship

20.    Paul, a 40-year-old father dying of cancer, makes frequent nasty remarks to his
       wife such as, "you can't wait until I die so you can get the insurance money and
       live it up, can you."  This illustrates which of Kübler-Ross's stages?
       a.    anger
       b.    bargaining
       c.    denial
       d.    depression

21.    Which of the following would Kohlberg say is typical of the most sophisticated
       level (stage 6, postconventional) of moral reasoning?
       a.    concerns for societal standards
       b.    the immediate consequences of one's actions
       c.    the need to maintain social order
       d.    the use of universal ethical principles, such as justice and human rights

22.    Which of the following is NOT a phase of grieving according to Glick, Weiss,
       and Parkes?
       a.    coping with anxiety and fear
       b.    initial response
       c.    mania
       d.    recovery

23.    Parents often feel that, as their child gets older, the child shares less intimacy
       with them.  Which statement about the intimacy between parents and
       adolescents is true?
       a.    adolescents become less intimate with peers during this period
       b.    adolescence is noted for a decrease in parent/child intimacy
       c.    adolescence is noted for an increase in parent/child intimacy
       d.    the intimacy the child has with the parent stays about the same through
             adolescence as it was in childhood

24. William Bennett, a former Secretary of Education, has said that adults should not make excuses for students when they steal; that it doesn't matter if the thief is poor or that the owner can easily replace the stolen object. Stealing is wrong. Mr. Bennett's remarks are most directly concerned with:
   a. cognitive learning
   b. developmental tasks
   c. ethical considerations
   d. morality

25. How do feelings about one's appearance relate to time of maturation for each sex?
   a. both early maturing boys and early maturing girls feel prouder of their appearance and more self-confident than their late maturing companions
   b. both early maturing boys and early maturing girls feel more embarrassed about their appearance and less confident than their late companions
   c. early maturing boys tend to feel embarrassed about their appearance, whereas early maturing girls are proud of their appearance
   d. early maturing boys tend to be more pleased with their appearance than late maturing boys, but early maturing girls tend to feel embarrassed about their appearance

26. How did Erikson identify the conflict experienced in adolescence?
   a. generativity vs. stagnation
   b. identity vs. role confusion
   c. integrity vs. despair
   d. intimacy vs. isolation

27. According to Nydegger and Mittness (1979), men who wait to have children may experience _____ conflict between family life and career demands.
   a. constant
   b. less
   c. more
   d. severe

28. {T F} The theory of moral development was developed by Erik Erikson.

29. Jan has been diagnosed as having widespread and inoperable cancer of the lungs. When told this, she said, "The lab must have gotten my X-rays mixed up with somebody else's--I never smoked that much!" Which one of Kübler-Ross's stages is she in?
   a. anger
   b. bargaining
   c. denial
   d. depression

30. The relationship between two unmarried people who live together is called

   _____.

31. Nita and Bill are 12-year-old twins. Which of the following is probably an accurate statement about their physical development?
   a. Nita may be taller and heavier than Bill.
   b. Nita should be heavier, but Bill should be taller
   c. Bill and Nita, because they are twins, should be about the same height and weight
   d. Bill may be taller and heavier than Nita

32. Which of the following is a primary sex characteristic?
   a. breasts of the female and large chest muscles of the male
   b. high voice of the female and deep voice of the male
   c. the ovaries of the female and the testes of the male
   d. wider hips than shoulders of the female and wider shoulders than hips of the male

33. _____ is the practice of ending a life for reasons of mercy.

34. According to Kohlberg, people reach the highest level of moral development before their twenties. One suggested way to facilitate attainment of this level is:
   a. being exposed to different cultures and value systems
   b. being exposed to the rules of various religions
   c. being taught democratic values that are universal
   d. to be taught to accept the behavior of others without judgment

35. The final stage of accepting one's impending death is, according to Kübler-Ross,
   a. acceptance
   b. despair
   c. isolation
   d. reactive depression

36. For Erikson, the opposite of intimacy is:
   a. feelings of inferiority
   b. isolation
   c. mistrust
   d. role confusion

37. According to Henning and Jardin (1976), women who had established themselves as successful executives by the end of young adulthood tended to:
   a. be the middle child
   b. have average college grades
   c. have extremely supportive fathers
   d. lack financial assistance

38. The life-span development approach has which of the following goals?
   a. describe and explain behavior changes within an individual and differences between individuals from conception to death
   b. describe the physical changes that occur between birth and adolescence
   c. explain the results of learning from conception to death
   d. investigate the influence of heredity and environment upon individuals from conception to death

39. {T  F} According to Levinson, there is a stable period enjoyed between 55 and 60 which he calls culmination of middle adulthood.

40. When does the adolescent growth spurt typically occur?
    a. between ages 10 and 13 for both sexes
    b. between ages 10 and 13 for boys, and 12 and 15 for girls
    c. between ages 10 and 13 for girls, and 12 and 15 for boys
    d. between ages 12 and 15 for both sexes

41. As young adults, late maturing males are:
    a. less likely to abuse alcohol than males who mature early
    b. more likely to abuse alcohol than males who mature early
    c. more likely to develop anorexia than early developing males
    d. more likely to have a higher body image satisfaction level than males who mature early

42. During which phase of the grief process is the person most likely to depend upon tranquilizers? The _____ phase.
    a. initial response
    b. recovery
    c. second
    d. third

43. Christine knows the doctors have given her only six more months to live, but she has been promising God she will stop smoking and drinking entirely if He will only giver her more time than that. Which one of Kübler-Ross's stages is she in?
    a. anger
    b. bargaining
    c. denial
    d. depression

44. Erikson felt that successful resolution of the midlife crisis would result in a person who is concerned with others and with improving society. He called this:
    a. generativity
    b. industry
    c. integrity
    d. intimacy

45. Kohlberg found that the typical responses of seven-year-old children to the moral dilemmas he presented them were at the _____ level.
    a. amoral
    b. conventional
    c. preconventional
    d. postconventional

## POSTTEST ANSWERS

1. T
2. b
3. b
4. d
5. c
6. c
7. b
8. menarche
9. b
10. F
11. b
12. d
13. b
14. d
15. c
16. d
17. c
18. a
19. d
20. a
21. d
22. c
23. d
24. d
25. d
26. b
27. b
28. F
29. c
30. cohabitation
31. a
32. c
33. Euthanasia
34. a
35. a
36. b
37. c
38. a
39. T
40. c
41. b
42. c
43. b
44. a
45. c

## PSYCH JOURNAL

**Please use the following pages to record your thoughts and feelings about the following questions.**

1. Write about the many things that determine happiness and satisfaction with your life, such as health, family, career, and financial security. Think about how these things changed throughout Georgia O'Keeffe's life, and write about how they might change throughout your life. At what age do you expect your overall happiness and satisfaction to be highest?

_____

_____

_____

_____

_____

_____

_____

_____

_____

_____

_____

_____

_____

_____

_____

_____

_____

_____

2. Discuss your own moral development and internalization; how did you acquire the values you hold most important today?

_____

_____

_____

_____

_____

_____

_____

_____

_____

_____

_____

_____

_____

_____

_____

_____

_____

_____

_____

_____

3.	Describe your own "identity crises."  What are the major conflicts between you and your parents?  How can you see these conflicts as evidence of your attempts to balance dependence and independence in your relationship with your parents?

_____

_____

_____

_____

_____

_____

_____

_____

_____

_____

_____

_____

_____

_____

_____

_____

_____

_____

_____

_____

4. Describe where each member of your family is in terms of their lifespan crises and issues. How might you be able to help family members make necessary transitions?

_____

_____

_____

_____

_____

_____

_____

_____

_____

_____

_____

_____

_____

_____

_____

_____

_____

_____

_____

_____

# Chapter 10 -- Motivation and Emotion

## CHAPTER SUMMARY

1.  Motives energize and direct the behavior of organisms; they include the goal toward which our behaviors are aimed. Primary motives concern our biological needs. Social motives come from learning and social interactions. Instinct theories argue that motivation is innate and unlearned. Drive theories state that we maintain an internal homeostatic balance. When that balance is upset, a need results. The need sets up a drive to correct the imbalance. Incentive theories focus on the role of external stimuli in "pulling" behavior. Cognitive theories suggest that behavior can be guided and directed by plans and goals developed by the individual.

2.  A system of glucoreceptors, located in different parts of the body, monitors the level of glucose in the blood and signals hunger and satiation. Taste also influences how much we eat. According to set-point theory, the number and size of fat cells in your body determine your weight. The number of fat cells is determined in early childhood. Learning affects what we choose to eat. Anorexia nervosa and bulimia are most common in women, from adolescence through early adulthood. Theories about the causes of these disorders characterize the individuals as suffering from a hormonal imbalance, trying to overcome depression or feelings of helplessness by exerting control over their environment, or being overly influenced by our society's emphasis on thinness.

3.  Osmometric thirst occurs when the intracellular water level is depleted. Volumetric thirst is experienced when fluid from extracellular areas is lost because of blood loss, diarrhea, or vomiting.

4.  The biological basis of male and female sexual behavior is hormonal (androgen in the male, estrogen in the female.) Masters and Johnson have categorized the human sexual response in four phases: the excitement phase, the plateau phase, the orgasm phase, and the resolution phase. Males then enter a refractory phase. Males generally have only one orgasm; woman may have multiple orgasms. Overall surveys indicate that we are becoming more sexually active at increasingly earlier ages. However, there are indications that because of the fear of AIDS, the number of sexual partners for both men and women is declining. Evidence suggests that although only a small portion of the population is exclusively homosexual, at least 20 percent of American adult men have had homosexual contact. This and other findings indicate that we should view sexual orientation on a continuum running from exclusive homosexual to exclusive heterosexual. Factors cited as possible causes for homosexual behavior include hormonal imbalances, genetic predisposition, and social influences during the period when the sex drive develops. Both sexual gratification and violence motivate rape. In some cases, the rapists wants to hurt and degrade the victim; in other cases, the aim is sexual gratification. Date rape is the term used to describe rape that occurs in the context of a dating relationship or between people who are acquainted.

5.  Stimulus-seeking motives do not seem to be caused by some internal state or need. Rather, these responses are elicited by new and complex stimuli. Also, the aim of these motives is to increase arousal or tension, not to reduce it.

6.      Frustration-aggression theory states that frustration instigates people to aggression.  After people act aggressively, they experience a release of tension called catharsis.  Attempts to modify frustration-aggression theory includes the theory that only frustration accompanied by anger leads to aggression.  Social learning theory states that we learn how to be aggressive by watching aggressive models and when aggressive actions are rewarded.  Research on media violence suggests that watching violence on television or in films increases the likelihood of engaging in aggression.  Violent pornography may also lead to more aggressive behavior toward women.  People high in achievement motivation (nAch) are driven by a desire to set high standards for excellence and to achieve goals.  Both learning goals and performance goals may be involved.  The motivation to achieve seems to develop early in life and is affected by family situations that encourage and reward independent behavior.  A task performed because it is enjoyable or meets internal standards is based on intrinsic motivation.  When a task is performed based on external rewards, it is an example of extrinsic motivation.  Intrinsic motivation can be changed into extrinsic motivation by offering high rewards for performing a task that was once enjoyable.

7.      Emotions are affective states (or feelings) usually instigated by external events accompanied by physiological changes that often influence behavior.  Moods are viewed as general feeling states that are less intense, less specific, and longer lasting than emotions.

8.      According to the James-Lange theory, external events cause physiological changes in people, and it is the perception of these changes that leads to emotions. Cannon argued that physiological changes occur too slowly to be the basis of emotion.  According to the Cannon-Bard theory, emotional experience and physiological arousal occur at the same time.  Cognitive theories emphasize the role of cognition or interpretation on emotion.  Emotions arise out of intuitive or automatic appraisal of events.  The somatic theory suggests that emotions are a result of labeling facial muscle reactions to events.  According to the opponent-process theory of emotion, each emotional state triggers a force to experience the opposite emotional state.  The effect is to dampen the intensity and length of time that we experience any emotional state, which may be very adaptive.

9.      Facial expressions communicate emotions.  Personal space is linked to culture and type of communication.  We move closer to people we like and away from those we dislike.  It is difficult to control our facial expressions and body language when we lie.  Thus, there may be a direct, possibly innate, link between feelings and expression.

10.      Love is characterized as a personalized form of joy that involves intimacy, passion, a commitment, and sacrifice.  The quality and intensity of love varies over time.  It has been suggested that people must learn to love and that our ability to love is affected by early attachment relationships.

# KEY TERMS AND CONCEPTS

Motivation:  History and Scope
       Motives energize and direct the behavior of organisms
           Primary motives - biological needs
           Social motives - learning and social interactions
       Instinct theories argue motivation is innate and unlearned
           Incorporated in ethology, sociobiology
       Drive theories state that we maintain an internal homeostatic balance
           Homeostasis - body wants to maintain a constant internal state
           Needs - result when homeostatic balance is upset
       Incentive theories focus on role of external stimuli in "pulling" behavior
       Cognitive theories suggest behavior can be guided by plans and goals

Hunger and Eating
       Biological foundations of hunger
           Gut reaction - cramping and pangs coming from an empty stomach
           Blood-sugar level - signal hunger and satiation
           Role of taste - influences how much we eat
           Body weight and long-term eating
       What we eat: specific hungers and learning
       Obesity
           Set-point theory - number of fat cells determine weight
           Influence of external cues affects what we choose to eat
       Anorexia Nervosa and Bulimia
           Most common in women from adolescence to adulthood
           Possible causes - hormonal imbalances, depression, and society's
               emphasis on thinness

Thirst
       Osmometric thirst - occurs when intracellular water is depleted
       Volumetric thirst - fluid is lost from extracellular areas because of blood loss

Sexual Behavior
       Biological basis of sex - hormonal (androgen in male, estrogen in female)
           Male biological sexual behavior
               Androgens sent to the bloodstream which increases sex drive
           Female biological sexual behavior
               Estrogen is secreted more during ovulation
           Importance of external cues
               Releasing stimulus - event that causes a predetermined response
       The human sexual response in four stages
           Excitement phase
           Plateau phase
           Orgasm phase
           Resolution phase
           Refractory phase (males)
       Human sexual behaviors
           Autosexual behavior - first and most common experience is masturbation
           Heterosexual behavior
           Homosexual behavior - sexual desire for those of the same sex
       Coercive Sex
           Rape and date rape - motivated by sexual gratification and violence
           Impact is severe and wide ranging for the victim

Stimulus-Seeking Motives
        Not caused by some internal state or need
        Elicited by new and complex stimuli
                Aim is to increase arousal or tension, not reduce it

Social Motives
        Aggression and violence
                Instinct theories - aggression is an instinct
                Biological theories - genes determine aggression
                Frustration-aggression theory
                        Frustration instigates people to aggression
                                Displaced aggression
                        Catharsis - release of tension
                Social learning theory
                        Learn to be aggressive by watching models
                Media violence and aggression
                Pornography and aggression
        Achievement motive
                Performance goals
                Intrinsic and extrinsic motivation

Emotion
        Affective states usually instigated by external events
        Moods - general feeling states that are less intense than emotions

Getting that Feeling: Theories of Emotion
        James-Lange theory
                External events cause physiological changes in people
        Cannon-Bard theory
                Emotional experience and physiological arousal occur at the same time
        Cognitive theories
                Role of appraisal - is situation threatening or nonthreatening
        Somatic theory of emotions
                Labeling facial muscle reactions to events
        Opponent-process theory
                Emotional states trigger a force to experience the opposite emotion

Nonverbal communication: Saying It Silently
        Face - communicates emotions
        Kinesis is the study of body language
                Personal space - the distance we put between ourselves and others
                        Intimate distance
                        Personal distance
                        Social distance
                        Public distance
        Emotional expression: innate or learned

Specific Emotions: A Focus on Love
        Love is characterized as a personal form of joy that involves intimacy
                Sternberg triangular theory of love
        Learning to love
                Affected by early attachment relationships
        Effects of arousal

## DISCUSSION QUESTIONS AND EXERCISES

**Note: These questions and exercises ARE the learning objectives for this chapter. Answer them accurately in your own words and you will have mastered the most important material. We guarantee it.**

1.  Motivation: History and Scope

a.  How have psychologists used the term **motive**? Define **motivation**. Contrast **primary motives** and **social motives**.

b.  The nature of the human organism has been questioned and debated. What is the difference between determinism and rationalism?

c.  What is instinct theory? Identify some of the criticisms of early instinct theories.

d.    On what principle did Clark Hull base his comprehensive theory of motivation?
Define this principle, and give an example.

e.    What role does cognition play as a guide of behavior?  Describe some early and
recent theories.

2.    Hunger and Eating

a.    Explain how insulin levels are related to hunger.

b.    How does taste and body-weight affect why and what we eat?

c.       Why does learning play a role in influencing what we eat?  Does culture have anything to do with this?

d.       Define **obesity** and give a brief description of its social and economic impacts on people.  What are some of the possible causes and cues of obesity?

e.       Contrast **anorexia nervosa** and **bulimia**.  What are some of the possible causes of anorexia and bulimia?

3.    <u>Thirst</u>

a.       How is water distributed in our bodies?

b.      What is the difference between **osmometric** and **volumetric** thirst?

4.      Sexual Behavior

a.      Give a brief description how the sex drive is developed in both men and women.  Why is female sexual behavior more complicated?

b.      What role do external cues have in releasing sexual behavior?

c.      Characterize each of the following phases of sexual response:

        (1)     Excitement phase

        (2)     Plateau phase

        (3)     Orgasm phase

        (4)     Resolution phase

d.      Identify some of the major findings in research on heterosexual behavior. Define autosexual behavior.

e.      What is the history of homosexual behavior?  What are some of the theories regarding the causes of homosexual behavior?

f.      Define rape and date rape.  What is the difference between the two?  What is the impact of rape on the victim?  What motives are behind rape?

5.      Stimulus-Seeking Motives

a.      Why is the motive to explore and manipulate important?

b. Some investigators have suggested that there may be a particular motive for arousal. What is the name for this and why is this motive present?

6. Social Motives

a. How do instinct theory and biological theory differ on their views of aggression?

b. Define **frustration** and **displaced aggression**.

c. What does social learning theory have to say about aggression?

d.   Describe the role of media violence, pornography, and aggression. How do they affect our behavior?




e.   What are some of the family characteristics and lifestyles of individuals with high achievement motive (nAch)?






f.   Are you studying in school because of intrinsic or extrinsic motivation? Contrast both of these types of motivation.






7.   Emotion

a.   What have investigators said about the difference between emotion and motivation? How have they distinguished emotions from moods?

8.    Getting that Feeling:  Theories of Emotion

a.    Give a brief history of the study of emotions.  How do the early theories differ
      from the later theories?

b.    There are several theories of emotions, each one with a different theme.
      Characterize each of the following theories:

      (1)    James-Lange theory

      (2)    Cannon-Bard theory

      (3)    Cognitive theories

      (4)    Somatic theory of emotions

(5) Opponent-Process theory

9. <u>Nonverbal</u> <u>Communication:</u>  <u>Saying</u> <u>It</u> <u>Silently</u>

a. What method was used to identify the critical features of facial expression? Why was this important?

b. What is the study of **kinesis**?  Identify three areas of **personal space** and give a description of that space.

c. What are some of the other factors that influence the personal space we use? What are some other aspects of body language that communicate emotion?

d.     Are the expressions of emotion natural innate tendencies or are they learned? Can we "fake" emotional expressions?

10.    <u>Specific</u> <u>Emotions:</u>  <u>A</u> <u>Focus</u> <u>on</u> <u>Love</u>

a.     Can you define love?  What are some of the various components of love?

b.     How might predictability affect a relationship?

c.      What did Harlow find in his experiments with monkeys?  What has research found with regard to the attachment process and its affect on love?

d.      What are some of the effects of arousal?

## POSTTEST

1.      Women experience menopause:
    a.      in the mid-sixties
    b.      in their late forties or early fifties
    c.      when they are no longer interested in sex
    d.      which signals the end of their sexual drive

2.      Homeostasis refers to
    a.      a balanced state in the body's internal environment
    b.      a state of optimum stimulation either above or below equilibrium
    c.      realizing one's full potential
    d.      the mechanism by which unconscious needs are satisfied

3.      According to the James-Lange theory, the conscious experience of emotion _____ physiological arousal; according to the Cannon-Bard theory, the conscious experience of emotion _____ physiological arousal.
    a.      coincides with; precedes
    b.      follows; coincides with
    c.      follows; precedes
    d.      precedes; follows

4. The theory positing that emotions are accompanied by distinct facial expressions is:
   a. cognitive
   b. James-Lange
   c. opponent-process
   d. somatic

5. What hormone is responsible for the greater amount of facial hair in men than in women?
   a. epinephrine
   b. estrogen
   c. norepinephrine
   d. testosterone

6. The needs for affiliation, aggression, and achievement are considered _____ motives.
   a. primary
   b. secondary
   c. social
   d. success

7. {T F} Motivation and emotion are separate and independent processes.

8. What is meant by the word instinct?
   a. a genetically determined and fixed way of reacting to certain stimuli
   b. a learned way of reacting to certain stimuli that had been displayed so frequently that it had become automatic and was now displayed without thinking
   c. an impulse to act in a certain way in response to a specified bodily sensation
   d. an unconscious force that caused humans to act irrationally

9. The study of body language, or communication through body movement, is called _____.

10. The refractory period is:
    a. a time following female orgasm during which females are largely unresponsive to further stimulation
    b. a time following male orgasm during which males are largely unresponsive to further stimulation
    c. the time between initiation of intercourse and orgasm
    d. the time between orgasms in multiorgasmic women

11. Walter was frustrated at work today. He could not attack his supervisor for fear of being fired. He came home and kicked his cat and yelled at his wife. This behavior is called:
    a. aggression
    b. catharsis
    c. displaced aggression
    d. displaced violence

12. What kind of parent-child relationship seems best to foster high need-achievement children?
    a. low-achieving parents, because the children will strive to do better than such parents
    b. parents who are aloof and distant, because their children are motivated to do something that will make the parents notice them
    c. parents who give rewards for success but also set high standards for their children
    d. parents who set high standards but give little reward for success

13. Which of the following statements about whether emotions are innate or learned is accurate?
    a. Emotions are equally affected by one's genes and environment.
    b. Emotions are predominately innate.
    c. Emotions are predominately learned.
    d. Learning affects both our expression and ability to interpret emotional cues.

14. What brain cells are involved in controlling water-drinking behavior?
    a. hippocampal cells
    b. osmoreceptors
    c. reticular cells
    d. VMH cells

15. In considering the motivation for rape, it is more often:
    a. a desire for sexual contact
    b. an outlet for aggression
    c. because a woman says no when she means yes
    d. caused by women who like to be dominated

16. Research has shown that hunger:
    a. is affected by external cues in both obese and nonobese subjects
    b. is affected by external cues in nonobese, but not obese, subjects
    c. is affected by external cues in obese, but not nonobese, subjects
    d. is not affected by external cues

17. {T F} It has been shown in research that injections of the male sex hormone, testosterone, can decrease aggressiveness, and that drugs that reduce the level of this hormone can increase violent aggression in males.

18. The rationalistic view of human nature is to _____ as the deterministic view is to _____.
    a. free choices; mechanical reactions
    b. Darwin; Descartes
    c. Descartes; Darwin
    d. mechanical reactions; free choices

19. In drive theories, the source of motivation lies _____ the organism; in incentive theories, the source of motivation lies _____ the organism.
    a. inside; outside
    b. inside; inside
    c. outside; inside
    d. outside; outside

237

20. For men, sexual activity is more likely to be motivated by desire for
_____; for women; it's more likely to be motivated by desire for
_____.
   a.   expression of love and emotional commitment; physical gratification
   b.   physical gratification; expression of love and emotional commitment
   c.   neither a nor b; there is no evidence for gender difference in this area
   d.   a is true, except in the case of homosexuals

21. Social learning theorists would suggest which of the following to parents?
   a.   allow children to behave violently in controlled settings to decrease their
        violence elsewhere
   b.   don't let children watch television
   c.   let children watch violent television shows to release their hostility
   d.   monitor and control the television programs you allow your children to
        watch

22. If someone steps on your toes in a crowded bus and you believe it was an
accident due to the shoving and pushing of other passengers, you are attributing
it to a/an _____ cause
   a.   extrinsic
   b.   intrinsic
   c.   primary
   d.   secondary

23. Which theory of emotion states that unexplained arousal leads people to search
their environment to find a label for the arousal which then leads to a specific
emotion?
   a.   Arnold-Lazarus
   b.   Cannon-Bard
   c.   James-Lange
   d.   Schacter-Singer

24. What hormone is responsible for the female having rounder hips than males?
   a.   androgen
   b.   epinephrine
   c.   estrogen
   d.   norepinephrine

25. Volumetric thirst triggers the _____ to release _____.
   a.   gonads, testosterone
   b.   pancreas, insulin
   c.   pituitary gland, ADH
   d.   thyroid gland, thyroxin

26. What theory of emotion assumes we first evaluate the situation we are in and
then, if we judge it as important, arousal occurs?
   a.   appraisal
   b.   bodily feedback
   c.   opponent-process
   d.   preconscious processing

27. People may sometimes give themselves away through nonverbal cues when they are lying. This is called:
   a. nonverbal kinesis
   b. nonverbal leakage
   c. verbal kinesis
   d. verbal leakage

28. Who showed that drives come from tissue needs by having a colleague swallow a balloon that showed stomach contractions when the colleague reported feeling hungry?
   a. Cannon
   b. Hull
   c. Richter
   d. Schacter

29. Throughout development, humans show curiosity about their environment. For example, a child may take apart an alarm clock to see how it works, an adult may visit new areas to see what is there. These types of behaviors are best described as:
   a. achievement motivation
   b. arousal motivation
   c. stimulus-seeking motives
   d. thrill seeking

30. A chemical secreted by one animal that affects the behavior of another is called a/an:
   a. aphrodisiac
   b. drive inducer
   c. hormone
   d. pheromone

31. {T  F} Learning plays a vital role in influencing what we eat.

32. Primary motives are:
   a. cognitive
   b. learned
   c. socially determined
   d. unlearned

33. In the context of hunger regulation, to what does the concept of "set point" refer?
   a. a target weight maintained by one's metabolism and hunger regulation mechanisms
   b. another name for the ventromedial area of the hypothalamus
   c. the maximum weight a rat can gain when its ventromedial hypothalamus has been lesioned
   d. the point at which a dieter will lose control of his or her diet and begin to overeat

34. Happiness generates unhappiness. Fear generates elation. These two statements are reflective of the _____ theory of emotion.
    a. appraisal
    b. opponent-process
    c. proponent-process
    d. somatic

35. Which of the following statements about catharsis is the MOST valid?
    a. decreases the probability of aggression
    b. has no effect on future aggression
    c. may increase the probability of aggression
    d. should be encouraged

36. Homosexual behavior or preference seems to be related to:
    a. a particular area of the hypothalamus
    b. an excess in estrogen
    c. a deficit in testosterone
    d. psychological problems

37. {T F} The personal space which measures from touching to 18 inches and used for comforting, protecting, loving, and telling stories is the personal distance.

38. Which of the following represents the correct sequence of the phases of the human sexual response?
    a. excitement, plateau, orgasm, resolution
    b. excitement, plateau, resolution, orgasm
    c. plateau, excitement, orgasm, resolution
    d. plateau, excitement, resolution, orgasm

39. What is the main difference between a drive and an instinct?
    a. a drive does not specify a particular pattern of behavior
    b. only an instinct is determined genetically
    c. only an instinct is physiologically arousing
    d. only humans have drives

40. Which of the following statements about masturbation is TRUE?
    a. It can be safely stopped by instilling guilt.
    b. It is not typically recommended by therapists in treating sexual disorders.
    c. Masturbation causes genital warts.
    d. Masturbation is not physically harmful.

41. The ability to experience love appears to be strongly influenced by one's:
    a. ability to fantasize
    b. degree of attractiveness
    c. past experience
    d. sexual preference

42. "Inner" causes of behavior are called _____ motives
    a. extrinsic
    b. intrinsic
    c. primary
    d. secondary

43. Manipulations that decrease blood glucose level cause _____; manipulations that increase blood glucose level cause _____.
   a. a decrease in general arousal; an increase in general arousal
   b. a decrease in hunger; an increase in hunger
   c. an increase in general arousal; a decrease in general arousal
   d. an increase in hunger; a decrease in hunger

44. Being "pulled" to eating on viewing an attractive food advertisement is an example of which of the following approaches to weight gain?
   a. arousal
   b. external cue influence
   c. internal cue influence
   d. set-point theory

45. Anorexia nervosa is to _____ as bulimia is to _____.
   a. binge eating and purging; self-starvation
   b. females; males
   c. males; females
   d. self-starvation; binge eating and purging

46. The main difference in function between short-term and long-term mechanisms that control our eating behavior is
   a. short-term mechanisms are sensitive to body weight while long-term mechanisms are focused on nutritional needs
   b. short-term mechanisms are primarily responsible for nutritional needs while long-term mechanisms are centered on skeletal development
   c. short-term mechanisms are primarily responsible for body weight while long-term mechanisms are responsible for energy allocation
   d. short-term mechanisms are primarily responsible for nutritional needs while long-term mechanisms are sensitive to body weight

47. One of the differences between male and female sexual response is that
   a. females show different changes in the breathing, muscle tension, and heart rate
   b. male genitals become enlarged whereas the female's remain the same size
   c. males have more variability in their orgasms than do females
   d. females have more variability in their orgasms than do males

48. Sally's boss, Steve, is often making unwanted passes at her and often telling her sexual jokes which offend Sally and make her uncomfortable. As a result, Sally no longer enjoys her job or likes coming to work. Steve's behavior could be characterized as:
   a. rape
   b. sexual harassment
   c. date rape
   d. sexual abuse

49. Seth walks into a room and people in party hats jump out at him and yell "Surprise!" This event stimulates the thalamus which in turn makes Seth's pulse race while simultaneously he experiences the emotion of surprise. These series of events can best be explained by which theory of emotion?
    a. James-Lange Theory of Emotion
    b. Cannon-Bard Theory of Emotion
    c. Schachter-Singer Theory of Emotion
    d. The Appraisal Theory

50. The assumption underlying the use of polygraphs is that the
    a. physiological response when lying is different than when telling the truth
    b. nonverbal behavior when lying is different than when telling the truth
    c. amount of a person's voice tremor can indicate lying
    d. reaction time to answer questions is different when a person is lying

## POSTTEST ANSWERS

1. b
2. a
3. b
4. d
5. d
6. c
7. F
8. c
9. kinesis
10. b
11. c
12. c
13. d
14. b
15. b
16. a
17. F
18. a
19. a
20. b
21. d
22. a
23. d
24. c
25. c
26. a
27. b
28. a
29. c
30. d
31. T
32. d
33. a
34. b
35. c
36. a
37. F
38. a
39. a
40. d
41. c
42. b
43. d
44. b
45. d
46. d
47. d
48. b
49. b
50. a

**PSYCH JOURNAL**

**Please use the following pages to record your thoughts and feelings about the following questions.**

1.  People who read about the experiences of the survivors of the crash in the Andes have a variety of emotional reactions. What were your feelings as you read of the adventure? Which theory of emotion do you believe best explains your emotions?

_____

_____

_____

_____

_____

_____

_____

_____

_____

_____

_____

_____

_____

_____

_____

_____

_____

2. The survivors of the crash took some drastic steps to keep alive, but many of the actions that violated social norms were taken with great pain and agony. As you examine the survivors' behaviors, consider what they suggest about the foundations of motivation. Think especially about the role of instinct, learning, and cognition in motivation. Have you ever made a difficult decision that called into question social norms? What was your motivation? What role did your emotions play?

_____

_____

_____

_____

_____

_____

_____

_____

_____

_____

_____

_____

_____

_____

_____

_____

_____

_____

_____

3. Do you often crave certain foods? Can you think of any nutritional element that these foods give you?

_____

_____

_____

_____

_____

_____

_____

_____

_____

_____

_____

_____

_____

_____

_____

_____

_____

_____

_____

_____

_____

_____

4. How did your family relate to food as you were growing up? Was food associated with emotional or social events? Have you found a need to re-think those habits now that you are away from home?

_____

_____

_____

_____

_____

_____

_____

_____

_____

_____

_____

_____

_____

_____

_____

_____

_____

_____

_____

_____

_____

5. Have you ever had the experience of crossed signals with a person you were dating? Why do you think date rape happens so often?

_____

_____

_____

_____

_____

_____

_____

_____

_____

_____

_____

_____

_____

_____

_____

_____

_____

_____

_____

_____

_____

_____

_____

6. In what situations do you behave aggressively or feel frustrated? How do you deal with this aggression or frustration? Is there a better way to deal with your feelings?

_____

_____

_____

_____

_____

_____

_____

_____

_____

_____

_____

_____

_____

_____

_____

_____

_____

_____

_____

_____

7.  Where do you side on the violence-on-T.V. issue? Do you remember
    mimicking actions from your favorite shows as a kid? When did you realize the
    difference between fantasy and reality? How did you learn that difference?

_____

_____

_____

_____

_____

_____

_____

_____

_____

_____

_____

_____

_____

_____

_____

_____

_____

_____

8. Which means more to you: to write a paper on a topic that interests you for which you get a "C" but learn a lot, or to write about a familiar, safe topic in a style that you know will get you a high grade? What is more important to you in your future career: high pay or some other factor?

_____

_____

_____

_____

_____

_____

_____

_____

_____

_____

_____

_____

_____

_____

_____

_____

_____

_____

_____

9. How sensitive are you to other people's body language? Can you give an example of a time when someone's body language contradicted their verbal message to you? Which message would you more likely believe?

_____

_____

_____

_____

_____

_____

_____

_____

_____

_____

_____

_____

_____

_____

_____

_____

_____

_____

_____

_____

# Chapter 11 -- Health Psychology

**CHAPTER SUMMARY**

1.     Since Freud's work with hysterics, the increasing awareness that emotions and behavior play a major role in influencing physical health, even when medical causes can be identified, has served as a foundation for **health psychology**. Soldiers returning from World War I suffered from a variety of illnesses with no clear physical cause.  Their cases enlarged the scope of health psychology.  The greatest growth in the field occurred when it embraced **wellness** and began to focus on improving lifestyles.

2.     **Stress** is the process by which people respond to events that are perceived as threatening or challenging.  **General adaptation syndrome**, the body's response to stress, involves three stages:  alarm reaction, stage of resistance, and the stage of exhaustion.  There is "good" stress and "bad" stress.  Change, whether it involves major life events or hassles, and unpredictability create stress.  Events that underscore our vulnerability such as earthquakes cause stress.  Also a perceived lack of control can contribute to stress.  Conflict occurs each time we make a choice, even when the choice is between two positive alternatives.  Environmental influences such as crowding, noise, and pollution cause stress.

3.     **Personality** influences how we handle stress.  People who exhibit Type A behavior with achievement strivings are not prone to illness.  However, people with Type A behavior who also show intolerance, anger, hostility, and an obsession with time experience higher levels of illness, especially heart attacks and depression.  People with hardy personalities tend to have a lower risk of illness.  They view situations as challenges rather than threats.  Culture influences our stress level by affecting our expectations about what events should take place, our interpretations of events, and the level of social support available to us.  People with low socioeconomic status experience more environmental stress, less control over their lives, and higher rates of mental illness than their wealthier counterparts.

4.     Our responses to stress are influenced by culture, learning, and genetics. A wide variety of illnesses including cancer and heart disease are linked to stress because stress reduces the body's immune system and often results in people neglecting their health.  **Learned helplessness** results when people believe that their actions have no effect on their outcomes.  Psychological disorders such as post-traumatic stress disorder are a result of stress.

5.     **Coping strategies** can fall into two categories:  problem-focused coping and emotion-focused coping.  Through reappraisal we can identify the source of the stress, reexamine the situation, and find a new perspective and new ways to deal with the stress.  Stress can be reduced by increasing people's belief in their self-efficacy. Getting information decreases uncertainty and helps develop an effective response. Social support involving a give-and-take relationship helps alleviate stress.  Social support can take many forms including physical assistance, material aid, intimate interactions, guidance or providing information, feedback, and social participation. Women generally have a stronger social support system and make better use of their support.  Changing our body's reaction through techniques such as meditation and biofeedback can help us cope with stress.

6.      We can take steps to improve our health by shaping our **lifestyles**. Smoking increases the likelihood of heart disease, lung cancer, stroke, and pneumonia. Passive smoke is also dangerous.  The health of people who quit improves dramatically.  Diets high in fat, refined sugar, and cholesterol and low in fiber not only increase the chance of heart disease, but they have also been linked to colon and rectal cancer.  Developing a good diet is important for both children and adults.  Sexual activity increases the risk of contracting a sexually transmitted disease (STD) such as gonorrhea, syphilis, herpes genitalis, and AIDS.  AIDS, caused by the HIV virus, destroys the body's immune system so that the individual is vulnerable to germs the body would normally destroy.  AIDS is presently incurable and terminal.  AIDS can be prevented in many ways including abstaining from sexual activity and intravenous drug use or using a condom during sexual intercourse.  Most importantly, know your sex partner and discuss AIDS issues with him or her.

7.      A regular program of **exercise** strengthens muscles, reduces weight, and has a positive influence on mental attitude.  Exercise need not be strenuous or painful to have a positive effect.  Patients must understand and respect their physicians so that they will follow their medical advice.  Physicians need to be sensitive to their patients' needs and cultural influences in order to provide better communication and treatment.


**KEY TERMS AND CONCEPTS**

The History and Scope of Health Psychology
    Sigmund Freud - hysterics (physical symptoms without physical cause)
    World War I and "battle fatigue"
    The scope of health psychology expanded in the 1960s
    Greatest growth - "wellness" and the identification of a healthy lifestyles

Identifying Stress and Its Origins
    Stress - response to events that are perceived as threatening or challenging
        Positive situations may also cause stress
        Stressors - events that give rise to stress
    General adaptation syndrome - Hans Selye's three phases
        Alarm reaction - body prepares to meet the stress
        Stage of resistance - bring emergency resources to a normal level
        Stage of exhaustion - body begins to wear down
    "Good" and "bad" stress - Selye and Lazarus
        Good stress - positive feelings about difficult demands
        Bad stress - negative feelings and damage
    Causes of Stress
        Change - anything from rare events to everyday events
            Two types of daily events
                Hassles - minor but daily annoying events
                Uplifts - completing tasks, feeling healthy
        Unpredictability - events you cannot plan for
        Lack of control - a belief or perception of no control
        Conflict - a person must choose between two courses of action
            Four major types of conflict
                Approach-approach conflict
                Avoidance-avoidance conflict
                Approach-avoidance conflict
                Double approach-avoidance conflict

Goals and regrets
Environment

Moderators of Stress
    Individual differences
    Personality styles
        Type A behavior - a response that involves becoming more active
            Two components
                Achievement strivings (AS) - taking one's work seriously
                Type A(II) - stress leads to an illness
            Hardiness - how people perceive stressful situations
                Challenged
                Committed
                In control
    Culture - establishes many expectations about what events should occur
    Social and economic class - influences the types of stressors

Negative Reactions to Stress - wide variations
    Illness - stress is a major factor in all types
        Attacks the immune system
        Physical damage may result (e.g., ulcers and heart disease)
    Giving up, or learned helplessness - the feeling of no control
    Psychological disorders

Coping Effectively with Stress - two categories (Lazarus)
    Problem-focused coping - identifying sources and reducing them
    Emotion-focused coping - changing emotions instigated by stress
        Reappraisal - taking a step back, or putting stress "on hold"
            What is creating the stress and changing your thinking
            Re-examine the situation and your ability to cope with it
        Belief in self-efficacy - judging our ability to cope with situations
        Getting information - a lack of information results in uncertainty
        Social support - network on whom you can rely in times of crisis
            Give-and-take relationship
            Many forms of social support
                Physical assistance
                Material aid
                Feedback
        Altering bodily reactions - techniques useful in reducing stress
            Meditation - exercises to gain greater control over mind and body
            Biofeedback - awareness of physiological measures (e.g., pulse)

Lifestyles and Health
    Smoking - smoking is linked with a variety of serious illnesses
    Diet - a critical factor in living a healthy life and avoiding illness
    Sexual behavior and sexually transmitted diseases (STDs)
        Gonorrhea - most common
        Syphilis
        Herpes genitalis - no known cure, but can be controlled
        AIDS - caused by HIV, no known cure

Promoting Healthy Behavior
    Exercise
    Physician-patient relationship

## DISCUSSION QUESTIONS AND EXERCISES

**Note: These questions and exercises ARE the learning objectives for this chapter. Answer them accurately in your own words and you will have mastered the most important material. We guarantee it.**

1.  The History and Scope of Health Psychology

a.   What is **health psychology**? From where does it get its roots?

b.   The greatest growth in health psychology occurred when it embraced the concept of **"wellness."** Where did this lead health psychology today?

2.   Identifying Stress and Its Origins

a.   Define **stress**. Name at least four things that can cause it.

b.   Define **stressor**. Give some examples of stressors in your life.

c.    Describe each of the following phases of a **stress response**. Identify the physiological reactions as well.

   (1)    alarm reaction



   (2)    stage of resistance



   (3)    stage of exhaustion



d.    Give some examples of "good" and "bad" stress as defined in the chapter.



e.    Define each of the following causes of stress and identify at least two characteristics.

   (1)    change



   (2)    unpredictability



   (3)    lack of control



   (4)    conflict

(5)     goals and regrets

(6)     environment

f.      What are **hassles**?  Give an example.

3.      <u>Moderators</u> of <u>Stress</u>

a.      Identify the two components of **Type A behavior**.  Which component is more
        likely to experience problems with stress?

b.      Describe the three Cs that compose the personality type labeled **hardy
        personality**.

c.       What role does culture play in our perceptions of stress?  Give some examples.

d.       How do social and economic class affect stress in our lives?  Give at least one example.

4.     <u>Negative</u> <u>Reactions</u> <u>to</u> <u>Stress</u>

a.       Identify some of the more general reactions to stress.

b.  What is **learned helplessness**?  What role does it play in reactions to stress?

5.   Coping Effectively with Stress

a.   Define the following two categories of dealing with stress, and give an example of each.

   (1)   problem-focused coping

   (2)   emotion-focused coping

b.   What is **reappraisal**? Identify the steps in the reappraisal process.

c.   Bandura suggested that stress could be reduced by increasing people's **belief in their self-efficacy**. What does this mean? Give an example.

d.   A lack of information will often result in a high degree of uncertainty, which is one of the major roots of stress. What can you do to avoid this problem?

e.   What role does social support play in coping with stress?

f.   Even if we cannot remove the source of stress, we have considerable control over how our body reacts to stress. What are some ways to help us control stress?

6. Lifestyles and Health

a.   Identify some behaviors that can decrease health and increase the chances of illness.

b.      Characterize each of the following **sexually transmitted diseases** (STDs):

   (1)    gonorrhea

   (2)    syphilis

   (3)    herpes genitalis

   (4)    AIDS

c.      Identify some of the things you can do to avoid sexually transmitted diseases.

7.    <u>Promoting</u> <u>Healthy</u> <u>Behavior</u>

a.    How does **exercise** help us when stress enters our lives?

b. Why is the physician-patient relationship important?

c. Identify some of the qualities that a physician should have to be effective in treating his or her patients.

## POSTTEST

1. Freud began his career by dealing with patients who were labeled as _____.
   a. dsyfunctional
   b. extroverts
   c. hysterics
   d. introverts
   e. psychoanalysts

2. {T F} There is an increasing realization that our behavior plays a more important role in influencing our physical health than do our emotions and feelings.

3. The field of psychology that examines how behavior affects physical well-being and contributes to the recovery from illness is called _____ psychology.
   a. community
   b. counseling
   c. environmental
   d. health
   e. social

4. {T F} The greatest growth in the field of health psychology occurred when it embraced the concept of the uniqueness of the individual.

5.  _____ is the process by which the individual responds to environmental and psychological events that are perceived as threatening or challenging.

6.  Events that give rise to stress are called _____.
    a.  challenges
    b.  hassles
    c.  releasers
    d.  stressors
    e.  threats

7.  {T F}  While it is possible for an event to be stressful for one person but not stressful for another, most stressful events are common across the vast majority of people in North America.

8.  Which of the following occurs during the process of appraisal?
    a.  After the initial shock wears off, we decide whether or not we are going to assess our ability to cope with it.
    b.  We judge the event as being harmful, threatening, or challenging.
    c.  We relax and deal with the problem accordingly.
    d.  We typically do all of the above, in the order given.
    e.  We typically do all of the above, but not necessarily in the order given.

9.  Which of the following is NOT one of the three phases of a stress response?
    a.  alarm response
    b.  appraisal
    c.  stage of exhaustion
    d.  stage of resistance

10. {T F}  The general adaptation response syndrome occurs more often with external stressors than it does with internal or emotional stressors.

11. There are two kinds of stress: _____ and _____.

12. {T F}  The range of events that gives rise to stress is almost unlimited.

13. Lisa was over 30 miles from home on her way to school when she realized that she had forgotten to feed Tom, her cat.  She was irritated and frustrated as she now had to drive all the way home to feed Tom.  This is an example that best meets the definition of a(n) _____.
    a.  bother
    b.  diversion
    c.  hassle
    d.  pain
    e.  uplift

14. Bob had three tests and a paper due in a three-day period.  He studied for three days in a row and hardly ate.  In what phase would he be near the end of this long three-day period?
    a.  alarm
    b.  appraisal
    c.  dsyfunction
    d.  exhaustion
    e.  resistance

15. Which of the following personality types has been correlated with heart disease?
    a.    Type A
    b.    Type A (AS)
    c.    Type A (I)
    d.    Type A (II)

16. What are the three Cs that appear in the personality style of hardiness?
    a.    challenged, committed, in control
    b.    challenged, composed, in control
    c.    competent, challenged, in control
    d.    composed, competent, courageous
    e.    composed, committed, courageous

17. {T F} Culture establishes many expectations about what events should occur.

18. {T F} Our culture influences how we will interpret events and, therefore, determines which situations will arouse stress.

19. {T F} The degree of comfort and support we receive from others when we are stressed or ill is influenced primarily by our personality.

20. The impact of class and economic condition is widespread, but it is seen most clearly in the area of _____.
    a.    education
    b.    health
    c.    housing
    d.    personality
    d.    technology

21. Of the following variables which does NOT influence our response to stress?
    a.    change
    b.    culture
    c.    history of learning
    d.    personality

22. Culture influences our reactions to stress because we _____ the responses that are acceptable and are likely to lead to attention from others within our culture.

23. Which of the following is NOT an example of stress in which actual physical damage results?
    a.    cancer
    b.    hair loss
    c.    heart disease
    d.    ulcers

24. Martin was recently diagnosed with a fatal disease. Instead of succumbing to depression and feelings of worthlessness, he decided to educate himself about the disease and change his lifestyle. How is Martin coping with this new stress?
    a.    He is using a defense mechanism to keep himself busy.
    b.    He is using the emotion-focused coping technique by changing his focus.
    c.    He is using the problem-focused technique in identifying his source of stress and taking action.
    d.    He will eventually stop avoiding his disease and deal with it.

25.     Mary is in the middle of a final exam in her economics class when her mind
        starts to wander and she begins to think about how this test will make or break
        her in business school.  If she doesn't get an A on this test she will flunk out of
        school.  She begins to panic.  In the midst of her darkest thoughts, she pauses
        and takes a step back, putting her stress "on hold".  What process is Mary using
        in this situation?
        a.      breaking the stressor
        b.      reappraisal
        c.      re-examination
        d.      retroactive analysis
        e.      seeking information

26.     Which of the following conditions is NOT associated with learned helplessness?
        a.      anger
        b.      apathy
        c.      depression
        d.      negative emotions
        e.      reduced intellectual abilities

27.     {T F}  Regular exercise may be helpful in reducing stress.

28.     {T F}  Searching for information typically helps you realize more and more
               about a situation, but in the process it also creates more stress with every
               new thing you learn.

29.     _____ _____ is a network of people on whom you can rely in
        times of crisis.

30.     The most positive type of social support in many situations involves a(n)
                    _____ relationship.
        a.      authoritarian
        b.      giving relationship
        c.      give-and-take relationship
        d.      helping relationship
        e.      submissive

31.     {T F}  Evidence suggests that talking with others about stress helps to reduce
               stress, as opposed to "holding it" inside.

32.     All of the following are forms of social support EXCEPT:
        a.      cultural diversity
        b.      feedback, such as helping one understand problems
        c.      guidance or providing information
        d.      intimate interactions
        e.      material aid

33.     {T F}  Even if we cannot remove the source of stress, we have considerable
               control over how our body reacts to it.

34.     {T F}  A healthy lifestyle will ensure that you seldom get sick.

35. Smoking has been linked to a variety of serious illnesses including all but which of the following:
   a. calcification of the bones
   b. emphysema
   c. lung cancer
   d. pneumonia
   e. stroke

36. {T F} There has been an overall increase in the number of people who smoke in the United States.

37. {T F} Diet is a critical factor in living a healthy life and avoiding illness.

38. _____ is an incurable virus, the symptoms of which include cold sores and fever blisters.

39. The most common STD in the United States, is
   a. AIDS
   b. gonorrhea
   c. herpes genitalis
   d. HIV
   e. syphilis

40. Which of the following STDs can be cured, but will lead to paralysis, or even death, if left untreated?
   a. AIDS
   b. gonorrhea
   c. herpes genitalis
   d. HIV
   e. syphilis

41. _____ is caused by the human immunodeficiency virus that destroys the body's immune system, causing the individual to become vulnerable to germs.

42. From what we currently know, how many years may it take before you begin to see the symptoms of AIDS?
   a. less than 1
   b. 2
   c. 5
   d. 7
   e. 10

43. {T F} AIDS can be avoided.

44. {T F} Exercise builds and strengthens muscles, reduces weight, and has positive mental effects because it allows the mind to focus on other activities.

45. For physicians to be effective in treating their patients, they must do which of the following?
   a. understand the importance of the relationship with their clients
   b. be competent
   c. be able to communicate their concern and advice to their patients
   d. all of the above

46. _____ conflict is the easiest to resolve while _____ is the most difficult to resolve.
    a.    approach-approach; approach-avoidance
    b.    approach-approach; avoidance-avoidance
    c.    approach-avoidance; double approach-avoidance
    d.    avoidance-avoidance; double approach-avoidance

47. Francis, an older woman, has a daughter Karen. According to the research on goals and regrets, Francis will:
    a.    view her regrets less negatively than Karen.
    b.    view her regrets more negatively than Karen.
    c.    view regrets in a more integrated, less stressful manner than Karen.
    d.    view regrets in a less integrated, less stressful manner than Karen.

48. Gaylon has a hardy personality type. As such, Gaylon is likely to:
    a.    let events control his life.
    b.    have very few regrets.
    c.    view stressful situations as a threat.
    d.    seek out help from others.

49. Socioeconomic position moderates stress in which two ways?
    a.    goals and regrets; amount of change
    b.    types of stressors; timing of stressors
    c.    amount of control; types of stressors
    d.    timing of stressors; amount of control

50. Which of the following people is most likely to suffer from peptic ulcers?
    a.    Angela, whose management position continually requires her to make important decisions
    b.    Ed, a college student who has not eaten well and has not exercised for a long time
    c.    Delia, whose job always requires her to respond quickly never knowing if she is making the right decision
    d.    Tony, who works a great deal around hazardous waste and is often aware of this danger

## POSTTEST ANSWERS

1.     c
2.     F
3.     c
4.     F
5.     Stress
6.     d
7.     F
8.     b
9.     b
10.     F
11.     good; bad
12.     T
13.     c
14.     d
15.     d
16.     a
17.     T
18.     T
19.     F
20.     b
21.     a
22.     learn
23.     b
24.     c
25.     b
26.     a
27.     T
28.     F
29.     social support
30.     c
31.     T
32.     a
33.     T
34.     F
35.     a
36.     F
37.     T
38.     Herpes genitalis
39.     b
40.     e
41.     AIDS
42.     d
43.     T
44.     T
45.     d
46.     b
47.     b
48.     d
49.     d
50.     c

# PSYCH JOURNAL

**Please use the following pages to record your thoughts and feelings about the following questions.**

1. When Gilda Radner learned she had cancer, her first question was, "Why me?" After reviewing Gilda's life and the material on health psychology, how would you answer this anguished question? What factors influence the likelihood of getting cancer?

_____

_____

_____

_____

_____

_____

_____

_____

_____

_____

_____

_____

_____

_____

_____

_____

_____

_____

_____

_____

2. It is never too late to develop a healthy lifestyle. Review your own life, and list the steps you should take to develop a healthier life style.

_____

_____

_____

_____

_____

_____

_____

_____

_____

_____

_____

_____

_____

_____

_____

_____

_____

_____

_____

_____

_____

3.    Describe your social support network.  Are there people to whom you could turn if you began to have problems?  How might you expand your network?

_____

_____

_____

_____

_____

_____

_____

_____

_____

_____

_____

_____

_____

_____

_____

_____

_____

_____

_____

# Chapter 12 -- Personality: Theories, Research, and Assessment

## CHAPTER SUMMARY

1.      Personality is defined as a unique set of enduring characteristics and patterns of behavior (including thoughts and emotions) that influence the way a person adjusts to his or her environment. The study of personality involves describing the structure of personality, identifying the dynamics or processes involved, and developing research tools and applications. Cultural factors need to be considered in the study of personality.

2.      People can be categorized as representing different types by analyzing traits. Traits are viewed as enduring qualities that cause people to act, think, or perceive in a relatively stable way across a variety of situations. The Big Five model of personality, comprising five traits, reduced personality into a manageable number of dimensions. Research, even across cultures, indicated that these five dimensions could predict a wide variety of behaviors. Eysenck identified three supertraits: extroversion, neuroticism, and psychotism. Eysenck also asserted that most of personality is inherited and that biological influences manifest themselves in brain function. Many personality theories assume that behavior is consistent. Research studies have shown that we can see general tendencies of stability and consistency over a broad range of time and situations, but we cannot make accurate predictions from one specific situation to the next.

Other research, however, has shown that situational variables may determine a person's behavior, overriding personality predispostions. Such factors as clarity and intensity of situational clues, how closely an individual pays attention to situational clues, and how important the personality trait is to the individual affect how a person behaves in any given situation. Historically, the notion that there are clear traits and behaviors distinguishing women from men was accepted. More recent research has identified masculine and feminine traits and evaluates the individual's androgyny. Researchers are also recognizing the influences of biology and culture on personality development. The trait approach has been criticized for leading to circular reasoning, including both description and evaluation, and not adequately explaining development and personality changes.

3.      Freud's psychoanalytic theory focuses on the internal, often unconscious forces that cause people to act in certain ways. Freud described personality as the result of a constant struggle between the id, ego, and superego. People are born with two innate drives: Eros (survival needs propelled by the libido) and Thanatos (destructive aims). Freud also identified specific stages of personality development (oral, anal, phallic, latency, and genital). An individual's personality is formed during these stages in early childhood. The neo-Freudians revised Freud's theory. Carl Jung disagreed with Freud's view of the libido as being the sexual energy of life. Jung developed a complex, mystical theory of personality that included the influence of the collective unconscious. He also believed that personality was not fully integrated until our middle and later years. Alfred Adler also disagreed with Freud's emphasis on sexuality. He proposed that people try to overcome their inferiorities by striving for superiority.

4.	Humanistic theories stress the creative aspect of people and argue that they are driven by the desire to reach their true potential.  Maslow suggested that human needs are arranged in a hierarchy like rungs on a ladder.  The most basic needs are on the bottom.  Self-actualization is at the top and the goal toward which we all strive.  Carl Rogers believed that people must accept themselves (their feelings and behaviors) before they can begin to reach their potential.  People can reach their potential only to the extent that they can accept all of their personal experiences as part of their self-concept.  Humanistic theories have led to research in such areas as self-esteem and creativity.

5.	According to the behaviorist approach, to predict how an individual will react in the present, one should examine that personal past schedule of reinforcement and punishment.  If we all grew up and existed in exactly the same environment, we would all behave the same way.  Social learning theories suggest that personality grows out of reciprocal interaction (the person and environment affect each other).  People acquire new behaviors through imitation.  Whether or not they perform these behaviors will be determined by the reinforcement contingencies available in the environment.  Cognitive learning theories suggest that thoughts can influence behavior.  Cognitive theorists have identified a variety of cognitions that affect behavior and explain individual differences.  Rotter believed that our general expectancies about our behavior and its rewards lead us to act consistently.  Another factor is self-efficacy (whether we believe we will be successful).  Mischel outlined five person variables (competencies, encoding strategies, expectancies, subjective values, self-regulatory systems and plans) that influence how individuals interpret and respond to certain situations.

6.	Personality assessment is the description and measurement of individual characteristics.  Behavioral assessment focuses on observing behavior in various situations.  Naturalistic observation allows us to examine people when they are unaware they are being studied.  Psychologists ask questions in structured and unstructured interviews.  Several questionnaires have been developed to examine specific parts of the personality.  The Minnesota Multiphasic Personality Inventory (MMPI) includes 550 true/false questions designed to measure ten different psychological disorders.  Projective tests are based on the psychoanalytic theory that unconscious motives influence behavior.  These tests ask people to give responses to ambiguous or vague stimuli.  The Rorschach Inkblot Test involves having people describe what they see in inkblots.  With the Thematic Apperception Test (TAT), people tell stories about what is happening in a series of pictures.


## KEY TERMS AND CONCEPTS

The History and Scope of Personality
	Personality
		A unique set of enduring characteristics and patterns
		Roots are in clinical psychology and social psychology
	Study of personality
		Describing the structure of personality
		Dynamics or process of personality
		Developing research tools and applications
	Personality in a cultural context
		"Nature versus nurture" -- How much personality is inherited
		Cultural factors need to be considered in study of personality

Structure of Personality: Types and Traits
        Describing personality is the major goal of the trait approach
        Traits are the building blocks of personality
                Cardinal traits - life revolves around these traits
                Central traits - guide a person's behavior
                Secondary traits - specific situations or events
        Genetics, biology, and trait theory
                The Big Five model
                        Most accepted model
                Eysenck's three supertraits
                        Extroversion, neuroticism, and psychotism
                        Two-thirds of personality can be traced to biological factors
        Research from the trait approach
                General tendencies of stability and consistency
                Gender differences in personality
        How consistent is behavior?
                Clarity and intensity of situational cues may be involved
                The individual is a factor in consistency
        Sex, Gender, and Personality
                Masculine and feminine traits are present
                Androgen levels predict some cognitive skills
                Culture expectations may vary for women and men
        Trait Approach: An Evaluation
                The approach has been criticized
                        Leads to circular reasoning
                        Not enough attention paid to influence of situation on behavior
                        Not adequately explaining development and personality changes

Theories of Personality: Dynamics and Process
        Freud's psychoanalytic theory
                Used the method of free association and dreams
                The conscious, preconscious, and unconscious
                        Anxiety leads us to use defense mechanisms
                Personality is the result of a constant struggle
                        Structure of personality
                                Id (savage) - satisfy our primitive desires
                                        Pleasure principle
                                        Instinctual drives
                                                Eros - drive for survival
                                                Thanatos - destructive drive
                                Ego - controls impulses of the id
                                        Reality principle
                                Superego - our conscious (ego ideal)
                Stages of personality development in early childhood
                        Oral stage - (1-2) focus on mouth region
                        Anal stage - (2-3) focus on anal area
                        Phallic stage - (4-6) discovery of genitals
                                Oedipus complex
                                Electra complex
                        Latency period - (6 - puberty) loss of interest in sex
                        Genital period - (puberty) sexual feelings reemerge
        Research from the psychoanalytic approach
                Some research has been used to support this theory
                Demonstration of defense mechanisms

Psychoanalytic theory: an evaluation
>Has influenced the study of human behavior
>Difficult to test many parts of psychoanalytic theory

Change of focus: the Neo-Freudians
>Carl Jung: Analytical Psychology
>>Disagreed with Freud's theory of libido
>>Jung had an optimistic view of human nature
>>Analytical psychology
>>>Ego, personal unconscious, and collective unconscious
>>>Personality not fully integrated until middle years
>Alfred Adler: Individual psychology
>>Disagreed with Freud's emphasis on sexuality
>>Inferiorities overcome by striving for superiority

Research from the Neo-Freudian approach
>Influence of birth order on personality and behavior

Humanistic Theories
>Stress the creative aspect of people
>>Argue that we are driven by the desire to reach our true potential

Abraham Maslow
>Human needs are arranged in a hierarchy
>Most basic needs are on the bottom, self-actualization is at the top

Carl Rogers
>People must accept themselves before they reach their potential
>Self-concept - judgments and attitudes about our behavior

Research from the humanistic approach
>Led to research in self-esteem and creativity

Humanistic theories: an evaluation
>Fail to identify the important terms of their theories
>Mostly concerned with human nature

Learning Theories
>Behaviorism
>>Predict how an individual will react in the present

Social Learning theory
>Personality grows out of reciprocal interaction
>>Person and the environment affect each other
>>Learning
>>>Imitation - watching others act, then copying it
>>Reinforcement and performance

Cognitive learning theory
>Thoughts influence behavior
>Locus of control -- Internals and externals
>Self-efficacy - degree of confidence we have in ourselves
>Mischel's five person variables that influence how we respond
>>Competencies - social skills and abilities
>>Encoding strategies - categorization of events
>>Expectancies - expectation in certain situations
>>Subjective values - value attached to various outcomes
>>Self-regulatory systems and plans - rules that guide behavior

Research from the learning and cognitive perspective
>Linking of cognition to motivation

Learning theories: an evaluation
>Behaviorism cannot explain the complex nature of human behaviors

Personality Assessment
    Description and measurement of individual characteristics
    Behavioral assessment
        Observing behavior in various situations
        Naturalistic observation
            People can be studied when they are unaware of it
    Interview
        Structured and unstructured
    Questionnaire
        Examine specific parts of the personality
            Minnesota Multiphasic Personality Inventory (MMPI)
    Projective tests
        Based on psychoanalytic theory
        Ask people to give responses to ambiguous or vague stimuli
            Rorschach Inkblot Test
                What people see in inkblots
            Thematic Apperception Test (TAT)
                Tell stories in a series of pictures

## DISCUSSION QUESTIONS AND EXERCISES

**Note: These questions and exercises ARE the learning objectives for this chapter. Answer them accurately in your own words and you will have mastered the most important material. We guarantee it.**

1.    The History and Scope of Personality

a.    Define **personality**. What is the goal of personality psychologists? Where are the roots of personality theory?

b.    Identify some of the different approaches of the study of personality.

c.     How does the western culture differ from other cultures such as Japan or India with regard to personality?

2.     <u>Structure of Personality: Types and Traits</u>

a.     What is the major goal of the trait approach to personality?  What are **types**? Give an example.

b.     Why is it not so easy to type people into distinct categories?  What approach was adopted with issues in consideration?  Give a description of this approach.

c.     Match the following trait with its appropriate description:

       1.     cardinal traits        _____ least influential characteristics
       2.     central traits         _____ guide a person's behavior
       3.     secondary traits       _____ traits which an individual's life revolves

d.	Most investigators argued that Allport identified too many traits.  Raymond eventually used factor analysis to identify clusters of traits.  What was the result of the intense effort to determine common clusters of trait descriptions?

e.	What was the impact of the Big Five model of personality?  What other model was offered?  What are the three supertraits?

f.	What type of studies have been conducted to determine how much of our personality is inherited?

g.	What ways did Eysenck suggest that biological influence occurs?  What has research shown with regard to extroverts and introverts?

h.    Under the influence of the trait approach what types of research have been conducted?

i.    Identify some of the different positions with regard to the consistency of behavior.  What factors may or may not influence behavioral consistency?

j.    Give a brief description of the various approaches to gender differences in personality.  What are some of the similarities and differences?  Which one do you agree with?

k.    What is the appealing nature of the trait approach?  Identify some of its major criticisms.

3.    <u>Theories</u> <u>of</u> <u>Personality:</u> <u>Dynamics</u> <u>and</u> <u>Process</u>

a.    What method did Sigmund Freud implement when he became dissatisfied with hypnosis?  Why?  Describe this method and how Freud used it.

b.    What two conclusions did Freud develop from his discussions with his patients?

c.    Characterize each of the following:

(1)    Conscious mind

(2)    Preconscious mind

(3)    Unconscious mind

d.  When we experience fear or anxiety because unacceptable impulses find their way into the conscious, it leads to defense mechanisms. What is the purpose of defense mechanisms?

e.  Briefly describe the three levels of personality listed below. State when, and by what process, each develops.

    (1)    id

    (2)    ego

    (3)    superego

f.  Characterize each of the five stages of personality development listed below. Identify how long each stage lasts, which levels of personality are present, and at least one major attribute of each stage.

    (1)    oral

    (2)    anal

    (3)    phallic

    (4)    latency

    (5)    genital

g.    What is the **Oedipus complex**?  How do young boys eventually resolve this complex?

h.    What is the **Electra complex**?  How do young girls eventually resolve this complex?

i.    What type of research has been drawn from the psychoanalytic approach?  What are some of the criticisms of Freud's theory?

j.    Some people found fault with the traditional psychoanalytic theory.  Carl Jung developed a complex, almost mystical, theory of personality.  What was this theory called?  What were Jung's three components?  Describe each one.

k.    What were Alfred Adler's view on Freud's theories?  How did he describe his approach?  What was the final version of his theory?

l.    What research was grown out of the neo-Freudian rebellion?

4.    <u>Humanistic</u> <u>Theories</u>

a.    Define **humanistic psychology**.  What were its goals?

b.    For each of the following theorists, give a description of his beliefs, major contributions, and goals.

      (1)    Abraham Maslow

(2)    Carl Rogers


c.    What is the **self-concept**?  What is your self-concept?


d.    Describe some of the humanistic approaches contributions to research.  What areas were studied?  What are some of the criticisms of the humanistic approach?


5.    Learning Theories

a.    What is the position of the learning theorists?  More recently, what is the focus of learning theory?


b.    Describe behaviorism in regard to personality theory.

c.	What is the position of social learning theories?  Why?  What are the two processes social learning theorists divide behavior into?

d.	Define both **imitation** and **observational learning**.  According to the social learning theory is reinforcement necessary for learning?

e.	Give a brief description of the cognitive theory of personality.

f.	According to Rotter (1966), we learn generalized expectancies to view reinforcing events either as being beyond our control or as being directly dependent on our actions.  What is the term used to describe this?  Contrast both internals and externals.

g.     Identify each of the five categories of cognitive factors that influence the way an individual interprets and responds to certain situations, as identified by Mischel.

h.     What is some research that has come from the learning and cognitive perspective?

i.     What are the strengths and weaknesses of the learning theories?

6. Personality Assessment

a. Define **personality assessment**. What is the aim of most assessment techniques? What methods are used to assess personality?

b. How is **behavioral assessment** used? Why? Identify and describe a method that is used in behavioral assessment.

c. What is an interview in personality assessment? How does a structured interview differ from an unstructured interview? How might the interviewer or the subject affect the validity of the interview?

d.  Why might a researcher use a questionnaire? Identify a personality questionnaire that is currently being used, and describe what is on this questionnaire.

e.  How have psychologists bypassed the gates that guard our unconscious? What are the two types of projective tests given in your text and describe their purposes?

## POSTTEST

1.  Sigmund Freud's method for the treating of disorders is called:
    a.  client-centered therapy
    b.  dream therapy
    c.  psychoanalysis
    d.  systematic desensitization

2.  According to Alfred Adler, the prime motivating force in a person's life is:
    a.  existential anxiety
    b.  physical gratification
    c.  striving for superiority
    d.  the need for power

3.    The pleasure principle is to the _____ as the reality principle is to the
      _____.
      a.    ego; id
      b.    ego; superego
      c.    id; ego
      d.    id; superego

4.    Awed by the grandeur of nature, the camper experienced a profound emotional
      high.  Maslow called such experiences:
      a.    archetypal experiences
      b.    ecstatic experiences
      c.    peak experiences
      d.    sublimation

5.    During the latency stage, children:
      a.    attempt to cope with the desires they have for their same-sex parent
      b.    begin to expand their social contacts beyond the immediate family
      c.    begin to focus their sexual energy on their opposite-sex peers
      d.    turn their biological urges loose

6.    Defense mechanisms combat feelings of anxiety and guilt:
      a.    by enhancing self-insight
      b.    by making unconscious urges conscious
      c.    through rational problem solving
      d.    through self-deception

7.    In explaining a person's aggressiveness, a social learning theorist would favor
      which of the following explanations?
      a.    an inferiority complex
      b.    observational learning
      c.    unconscious forces
      d.    unresolved conflicts

8.    {T  F} Because so much of Freud's theory seems difficult to support
            empirically, it has had very little impact on modern psychology.

9.    When we say that Tim is biosterous and happy, we are describing his:
      a.    perception of himself
      b.    personality traits
      c.    philosophy of life
      d.    unconscious motives

10.   In what system is "culture" a factor?
      a.    Cattell's 16 Personality Factors
      b.    Eysenck's two-dimensional model
      c.    Rogers's projective model
      d.    the "big five" personality factors

11.   The strongest support for the theory that personality is heavily influenced by
      genetics is provided by strong personality similarity between:
      a.    fraternal twins reared together
      b.    identical twins reared apart
      c.    identical twins reared together
      d.    nontwins reared together

12. To differentiate his approach from Freud's psychoanalytic theory, Carl Jung used the name:
   a. analytical psychology
   b. deterministic psychology
   c. existential psychology
   d. individual psychology

13. In order to identify clusters of closely related personality traits and the factors underling them, Raymond Cattell used:
   a. a factorial anova
   b. factor analysis
   c. the analysis of variance
   d. the normal distribution

14. Amy is torn between the need to study for an exam and her desire to go out with her friends. She decides that she will go out later only if she completes her studying. This realistic decision reflects the functioning of Amy's
   a. ego
   b. id
   c. superego
   d. unconscious

15. Projective tests such as the Thematic Apperception Test are designed to assess:
   a. the way others perceive you
   b. your behavior patterns
   c. your mental abilities
   d. your unconscious concerns, conflicts, and desires

16. Carl Rogers's view of personality structure centers around a single construct called:
   a. personality traits
   b. the ego
   c. the environment
   d. the self-concept

17. In Freudian terminology, the energy force that propels people to satisfy the drive for survival is called _____.

18. A Freudian might explain a compulsive smoker's behavior as being the result of fixation at the:
   a. anal stage
   b. genital stage
   c. latency stage
   d. oral stage

19. Operant conditioning is to B. F. Skinner as _____ is to Albert Bandura.
   a. classical conditioning
   b. genetics
   c. social learning theory
   d. the need for self-actualization

20. {T F} Personality psychologists are not only interested in how people are different from each other, but also in how they are similar to each other.

21.	Gordon Allport would call the compassion and altruism of Mother Teresa a
	_____ trait.
	a.	cardinal
	b.	central
	c.	secondary
	d.	surface

22.	What perspective on personality uses factor analysis?
	a.	humanistic
	b.	behavioral
	c.	psychoanalytic
	d.	trait

23.	Which of the following did Carl Rogers believe fosters a congruent
	self-concept?
	a.	appropriate role models
	b.	conditional positive regard
	c.	immediate-need gratification
	d.	unconditional positive regard

24.	In Freudian theory, the boy identifies with his father because of the
	_____ complex.

25.	Always having been a good student, Marvin is confident that he will do well in
	his psychology course.  According to Bandura's social learning theory, Marvin
	would be said to have:
	a.	a strong sense of superiority
	b.	strong defense mechanisms
	c.	strong feelings of self-efficacy
	d.	strong feelings of self-esteem

26.	Although Wilma is generally outgoing, she is very shy when she is with her
	close friends.  This tendency to be shy on occasion is an example of what
	Allport called:
	a.	cardinal traits
	b.	central traits
	c.	secondary traits
	d.	surface traits

27.	The most optimistic view of human nature is found in the:
	a.	behavioral approach
	b.	cognitive approach
	c.	humanistic approach
	d.	psychoanalytic approach

28.	_____ _____ is a Freudian method of getting patients
	to express every thought.

29. Mary's experiences have led her to believe that she is responsible for most of the things that happen to her. Julian Rotter would say that Mary tends to have:
   a. a strong superego
   b. an external locus of control
   c. an inner-directed self
   d. an internal locus of control

30. Barry sets extremely high standards for both himself and others. He tends to be rigid and inflexible and rarely allows himself to enjoy life. Freud would probably conclude that Barry is dominated by:
   a. his ego
   b. his id
   c. his superego
   d. the oral stage

31. A personality measure that asks you to respond freely to an ambiguous stimulus such as a picture is called a:
   a. behavior rating
   b. deceptive test
   c. projective test
   d. self-report personality inventory

32. {T F} The MMPI was originally developed to identify people with psychiatric disorders.

33. Both Carl Jung and Alfred Adler were especially critical of Freud's emphasis on:
   a. defense mechanisms
   b. sexuality
   c. the influence of childhood experiences
   d. the unconscious

34. During the _____ stage, girls discover they are inferior to boys and develop what Freud called the _____ complex.

35. According to Freud, harsh toilet training could well lead to fixation in the:
   a. anal stage
   b. genital stage
   c. oral stage
   d. phallic stage

36. Heather is an unfailingly polite person who always considers the feelings of others. This tendency to act in a similar manner across situations is indicative of which of the following qualities of personality?
   a. consistency
   b. distinctiveness
   c. reflexivity
   d. social competence

37. _____ refers to the extent to which people feel in control of their rewards and punishments.
    a. emotional stability
    b. locus of control
    c. reaction range
    d. trait centrality

38. Lisa is very quiet around people in authority, but she can be loud and boisterous among her peers. Which of the following theorists would explain the difference in Lisa's behavior in terms of situational forces?
    a. Alfred Adler
    b. Carl Rogers
    c. Gordon Allport
    d. Walter Mischel

39. {T  F} Even well-trained psychologists do a poor job of personality assessment if their judgments are based on a short interview.

40. Abraham Maslow called the need to fulfill one's potential the need for:
    a. achievement
    b. affiliation
    c. power
    d. self-actualization

41. Failure to resolve conflict at a particular stage of psychosexual development may lead to failure to move forward psychologically, a phenomenon that Freud called:
    a. compensation
    b. displacement
    c. fixation
    d. reciprocal determinism

42. Which of the following gives the stages of psychosexual development as outlined by Freud in their correct order?
    a. anal, latency, oral, phallic, genital
    b. anal, oral, latency, phallic, genital
    c. oral, anal, latency, phallic, genital
    d. oral, anal, phallic, latency, genital

43. The fact that Eysenck found introverts to be more easily conditioned than extroverts provides support for his theory that personality is influenced by:
    a. genetics
    b. the environment
    c. the unconscious
    d. whether one is introverted or extraverted

44. In explaining an individual's aggressiveness, Skinner would look for:
    a. an inadequate sense of self-worth
    b. early learning experiences and reinforcement history
    c. feelings of repressed hostility
    d. instinctual explanations

45. Billy has devoted his life to the search for physical pleasure and immediate need gratification. Freud would say that Billy is dominated by:
   a. his ego
   b. his id
   c. his superego
   d. his preconscious

# POSTTEST ANSWERS

1. c
2. c
3. c
4. c
5. b
6. d
7. b
8. F
9. b
10. d
11. b
12. a
13. b
14. a
15. d
16. d
17. libido
18. d
19. c
20. T
21. a
22. d
23. d
24. Oedipus
25. c
26. c
27. c
28. Free association
29. d
30. c
31. c
32. T
33. b
34. phallic; Electra
35. a
36. a
37. b
38. d
39. T
40. d
41. c
42. d
43. a
44. b
45. b

## PSYCH JOURNAL

**Please use the following pages to record your thoughts and feelings about the following questions.**

1.  Harry Truman was a complex individual. Historians have attempted to explain his rise to power and his success as president by focusing on the events of his time. Take the position of a personality psychologist. From what you have learned about personality theories, how would you explain Truman's personality? In other words, how have the personality theories helped you gain additional insight into Harry Truman? Take one of your favorite personality approaches and use it to explain Truman's actions.

_____

_____

_____

_____

_____

_____

_____

_____

_____

_____

_____

_____

_____

_____

_____

2.    Ask your parents whether they believe more in "nature" or "nurture" where you are concerned.  Did you arrive in the world with a definite personality, or did you develop traits more in response to your environment?  Are you a lot like one or both of your parents?  Your siblings?  To what degree?  What are your main differences?

_____

_____

_____

_____

_____

_____

_____

_____

_____

_____

_____

_____

_____

_____

_____

_____

_____

_____

_____

_____

3. Can you think of events that happened before you were four years old that likely shaped the kind of person you have become?

_____

_____

_____

_____

_____

_____

_____

_____

_____

_____

_____

_____

_____

_____

_____

_____

_____

_____

_____

_____

_____

4. Are there aspects of your personality that have developed because you are trying to make up for some kind of perceived lack? (Like becoming a physical fitness enthusiast because you were once considered a "98-lb. weakling".)

_____

_____

_____

_____

_____

_____

_____

_____

_____

_____

_____

_____

_____

_____

_____

_____

_____

_____

_____

5. Do you personally feel that people are inherently evil or tainted, or do you believe that most people strive to fulfill their loving potential?

_____

_____

_____

_____

_____

_____

_____

_____

_____

_____

_____

_____

_____

_____

_____

_____

_____

_____

_____

6.     What is your ethnic background? How do these ethnic groups fit and contribute in U.S. society? Do you know anyone from another culture who has found assimilation into U.S. society painful or difficult?

# Chapter 13 -- Abnormal Psychology

## CHAPTER SUMMARY

1.      To classify disorders, most mental health professionals use the *Diagnostic and Statistical Manual of Mental Disorders* now published in its fourth edition and referred to as DSM-IV.  Classification is based on a description of the disorder rather than its causes.  In DSM-IV individuals are evaluated along five different dimensions, a system described as multiaxial.  There is a growing recognition that culture plays an important role in the nature of disorders and the behaviors that are considered "abnormal."  Insanity is a legal classification to describe people whose mental condition makes them unable to understand inappropriate or illegal conduct.

Six models have been developed to identify the causes of abnormal behavior.  The medical model focuses on finding biological and organic causes of disorders.  The psychoanalytic approach argues that abnormal behavior is the result of unresolved conflicts that occur during infancy and childhood.  According to the learning/behavior perspective, maladaptive behavior is learned through the same principles of reinforcement and modeling that shape other behaviors.  The cognitive model finds the basis for irrational or bizarre behaviors in irrational beliefs and attitudes.  The humanists believe that disorders arise when people are prevented from satisfying their basic needs or are forced to live up to the expectations of others in their surroundings.  The systems model suggests that it is the social system (family, friends, community, society, and economics) that plays a major role in fostering maladaptive behavior in individuals.  To evaluate psychological disorders, many mental health professionals use a biopsychosocial orientation that assumes that human behavior is best understood as an interaction of biological, psychological, and sociocultural factors that contribute to both the psychological resources and psychological vulnerabilities of the individual.

2.      Anxiety disorders occur when people experience anxiety that is out of proportion to the situations they are in, and their responses to these anxiety-arousing situations are exaggerated and interfere with normal daily functioning.  Generalized anxiety disorder results when people are in a constant state of agitation or dread and cannot identify its source.  Panic disorder has a sudden onset of anxiety, overcoming the person with physiological symptoms.  Post-traumatic stress disorder involves reexperiencing traumatic events and the stress associated with them long after the events have passed.  Phobia occurs when people have an irrational level of fear of an object or event.  An obsessive-compulsive disorder is diagnosed when a person feels an overwhelming need to think a certain thought (obsession) or perform a certain behavior (compulsion).  People with somatoform disorders have physical complaints or symptoms that point to disease or damage to the body, yet there is no apparent organic basis for these symptoms.  Hypochondriasis involves a fear of disease or death.  People with somatization disorder complain of many symptoms and make frequent trips to doctors.  A person with conversion disorder reportedly has lost functioning in some part of his or her body, and there is no organic reason for it.

Dissociative disorders are characterized by the "loss" of some aspect of the personal experience of identity.  In the case of psychogenic amnesia, people forget past events or information, even their own identity.  Multiple personality disorder involves the development of two or more different personalities that are displayed at different times.  Mood disorders involves sadness and elation.  Depression is characterized by feelings of sadness, loneliness, boredom, and despair.  Depressed people often think of suicide. Depressed people may often neglect their appearance so they may look shabby. There are many theories about the causes of depression. Depression is a major factor in suicide. The best predictors of suicide are past suicide attempts and the threat of a suicide. The symptoms of mania include high energy levels, an inability to concentrate or stay with one task until it is completed, and disturbed sleep. In the case of bipolar disorder, a person alternates between mania and depression.

3.      Social disorders involve behaviors that inhibit developing satisfying social relationships. Sexual disorders involve sexual behaviors that cause the person distress, or hurt and debase others. Paraphilias fall into three different classes: sexual attraction to nonhuman objects; repeated sexual activities that hurt or humiliate the partner; and repeated sexual activity with children or nonconsenting adults. People who suffer sexual dysfunction are unable to perform or enjoy sexual intercourse. Sexual dysfunctions include sexual disorders and orgasm disorders. Substance use disorders comprise two categories: substance abuse (the pathological use of drug) and substance dependence (addiction including the presence of tolerance or withdrawal). Alcoholism is a form of drug addiction. Problem drinkers often drink too much, negatively affecting their lives and those of others. Alcoholics are addicted to alcohol. They have an uncontrollable urge to drink, and because of an increasing tolerance, they must continue to increase the dosage of alcohol. Drug addiction may cause physical dependence of psychological dependence. Drug addiction often has negative legal, economic, and social consequences. Personality disorders are rigid, maladaptive general approaches to dealing with events. People with personality disorders have problems with social interactions for which they blame others. Because they do not recognize their own problems, they seldom seek help. People with antisocial personality disorder perform violent and hurtful acts without the least bit of guilt, are incapable of forming close relationships, and are manipulative and insincere. They also may be charming and very intelligent. In addition, they often end up in prison.

4.      People suffering from psychosis have lost touch with reality. Psychotics create their own world. Schizophrenia, the most common psychosis, describes a group of disorders characterized by disorganization of thoughts, perceptions, communication, emotions, and motor activity. Schizophrenics may experience delusions, hallucinations, inappropriate emotions, and repetition of motor activities. Theories on the causes of schizophrenia abound. Genetics and the action of neurotransmitters may play an important role. Stressful family interactions may also contribute to the development of schizophrenia.

## KEY TERMS AND CONCEPTS

Abnormal Behavior: History and Scope
    The area of clinical psychology is devoted to this field
    Defining abnormality
        Labeled abnormal when behavior is unusual, causes distress
        Ways to define abnormality
            Statistical definition - as defined in the dictionary
            Cultural definition
            Maladjustment and personal distress - most important dimension
    Classifying disorders
        Emil Kraepelin listed 16 major categories of mental disorders
            Symptoms alone were inadequate for classification
            Symptoms must have three additional categories

Circumstances precipitating the disorder
The outcome of the disorder
The course of the disorder
This was the primary classification criterion
Today the *Diagnostic and Statistical Manual of Mental Disorders* is used
There have been four revisions of this manual
It is a multiaxial system
Goal - avoid speculation on causes; just describe
Axis I - major clinical syndrome
Axis II - developmental and personality disorders
Axis III - physical disorders or illnesses
Axis IV - psychosocial stressors
Axis V - level of adaptive functioning
This system is very complicated and requires training
Legal side: classifications by the courts
Judicial system has two concerns that require assessments of defendants
Competency to stand trial
A defendant must understand the charges against them
Does not absolve them of guilt
State of mind at the time of the crime
Does affect the person's guilt for that crime
Insanity is a legal term, not a psychological term
Identifying the causes of maladaptive behavior
Medical model
Psychological problems thought to result from disease
This model has considerable explanatory power
Psychoanalytic model
Unresolved conflicts during infancy or childhood
People experience anxiety and cope by using defense mechanisms
Psychosis may develop when people can't adjust
Learning/Behavioral model
Maladaptive behavior is learned, a product of the environment
Cognitive model
Behavior is strongly influenced by beliefs and attitudes
Change irrational behavior by changing attitudes
Humanistic model
Losing touch with one's feelings, goals, and perceptions
Systems model
Our social network has an impact on our behavior
Abnormal behavior may be caused by a "sick" system
Criticized because this model can only show relations not causes

Psychological disorders
Nearly half of all people in the U.S. experience a psychological disorder
Highest "at risk" groups
White women
People with incomes below $19,000
Those with fewer than 11 years of education
Individuals living in urban areas
Blacks are less likely to suffer most disorders than whites
Only 40% of those who have a disorder seek professional help

Emotional Disorders
  Anxiety disorders - most prevalent
    Generalized anxiety disorder and panic disorder
      A constant state of agitation and dread
      Symptoms
        Trouble falling asleep
        Loss of appetite
        Feel tense and keyed up
        Increased heart rate and faintness
      Panic disorder
        More bodily symptoms than generalized anxiety disorder
      Twice as common in females than in males and runs in families
    Post-traumatic Stress Disorder
      Reexperiencing traumatic events and the stress associated
      This disorder is not confined to wartime experiences
    Phobia
      Irrational level of fear to an object or event
      Controlled by avoiding the objects or events
      Most common in the United States
      Freud was one of the first to study phobias
    Obsessive-compulsive Disorder
      Obsession - recurring, irrational thought
      Compulsions - irrational behaviors or rituals
      Some explanations for obsessive-compulsive disorder
        Psychoanalytic theory
          Avoid threatening thoughts
          Guilt may underlie this disorder
        Cognitive theory
          Caused by faulty judgments about objects
  Somatoform disorders
    Physical complaints or symptoms that point to disease
      Hypochondriasis
        Fear of disease and death
        Reports of feelings of anxiety and depression
      Somatization disorder
        Focus on the symptoms themselves
        Feelings of anxiousness and depression
        Make frequent trips to multiple doctors
      Conversion disorder
        The loss of functioning in some part of the body
        Relatively easy to diagnose because of no organic damage
        May appear shortly after some stressful event
  Dissociative disorders
    The loss of certain psychological functions such as memory or identity
    Psychogenic Amnesia
      Memory loss due to extreme emotional distress or trauma
      People with this are unconcerned that they have forgotten
    Fugue States
      Loss of memory, and leaving suddenly for long periods of time
      Trading one memory loss for another
    Multiple Personality disorder
      One person assumes two or more personalities
      May develop because of severe and prolonged childhood trauma

Mood disorders
 Depression
  Loss of interest in most activities (eating, sex)
   Begins to affect the physical appearance of the person
  Thoughts of suicide are common
  Theories of depression
   Psychoanalytic theory
    Cause found in first mother-infant interaction
   Learning theory
    Few positive rewards are found in environment
   Cognitive theory
    Learning negative views of world and themselves
    Cognitive triad
     Negative thoughts about self, world, future
   Cognitive-learning approach
    Hopelessness model
   Medical model
    Emphasizes the physiological basis of depression
 Mania and Bipolar disorders
  Maniacs are a fountain of energy ready to take on the world
   Very positive self-image
   Easily distracted, increased sex drive, need less sleep
  Bipolar disorder
   Alternate between manic and depressive states
   Heredity plays a strong role in this disorder
 Suicide - the planned taking of one's life
  Depression plays a major role in 80% of suicide attempts
  There are sex differences in success of suicide
  Reactions to suicide
   Suicide is a mental disorder and a crime
   Suicide is a "fundamental right" of adults

Social Disorders - keep people from enjoying healthy, satisfying, social relationships
 Sexual disorders
  Paraphilias - sexual practices that fall outside of society's norms
   Sexual attractions to nonhuman objects
   Sexual activity which involves pain or humiliation
   Repeated sexual activity with children
  Sexual dysfunctions - unable to perform or enjoy sexual intercourse
   Sexual desire disorders - little or no desire for sex
   Orgasm disorders - recurring delay or absences of orgasm
 Substance Use disorder: Alcohol and other drugs
  Two categories: substance abuse and substance dependence
  Alcoholism - most common addiction in today's society
   Two categories of alcohol abuse
    Problem drinker - drinks too much
    Alcoholic - dependence on alcohol
  Alcoholism may be learned or inherited
 Drug addiction
  Different types of drugs
   Narcotics, stimulants, hallucinogens, or depressants
  Use of drugs has a long history
  Types of dependence

Physical dependence - body develops a need for drug
Psychological dependence - feel they must have drug
Personality disorders - rigid and maladaptive ways of dealing with environment
Seldom seek treatment
Antisocial personality disorder
Performance of violent and hurtful acts without guilt or regret

Psychotic Disorders
Schizophrenia - a group of disorders
Symptoms of schizophrenia
Disorganization of thought
Delusions - irrational beliefs
Disorganization of perception
Hallucination - sensory experience with no stimuli
Disturbances in communication
Inappropriate emotions
Unusual motor activities
Theories of schizophrenia - no specific cause has been identified
Biological/Medical models
Identical/Fraternal twin studies
Learning approaches
Develops as a result of an attempt to cope with stress
Cognitive theories
Lack ability to block out irrelevant stimuli

## DISCUSSION QUESTIONS AND EXERCISES

**Note: These questions and exercises ARE the learning objectives for this chapter. Answer them accurately in your own words and you will have mastered the most important material. We guarantee it.**

1.   Abnormal Behavior: History and Scope

a.   Define **abnormal behavior**. Identify and describe some of the various ways that abnormality can be defined.

b.     Attempts to classify psychological disorders can be traced to the psychiatrist
       Emil Kraepelin.  What was the strength of Kraeplin's classification system?
       What were the three categories Kraepelin insisted that symptoms had to be
       evaluated?

c.     Identify the classification system most mental health professionals use today.
       What system are individuals showing psychological distress evaluated on?

d.     For the following axes of the *Diagnostic and Statistical Manual of Mental
       Disorders* give a brief description of the information found in each axis:

       (1)     Axis I

       (2)     Axis II

       (3)     Axis III

       (4)     Axis IV

(5)     Axis V


e.      Describe the two concerns of the judicial system that require assessments of defendants.  What is insanity and how is it proven in a court of law?


f.      What are some of the other reasons for a rare plea of insanity?


g.      Describe how each of the following theoretical viewpoints (models) describe the mechanisms and processes that cause psychological disorders.  Identify some criticisms if there are any.

        (1)     Medical Model


        (2)     Psychoanalytic Model


        (3)     Learning/Behavioral Model

(4)     Cognitive Model

(5)     Humanistic Model

(6)     Systems Model

2.     <u>Psychological</u> <u>Disorders</u>

a.     Provide some of the various statistics on psychological disorders.  Who is the highest "at risk" group?  What are some other factors that are found in psychological disorders?

3.     <u>Emotional</u> <u>Disorders</u>

a.     What characteristics are common to the group of conditions classified as **anxiety disorders**?  What are **general anxiety disorders**?  How common are these disorders?

b.      What are some of the differences between **panic disorders** and general anxiety disorders?  What are some of the reasons why people suffer these disorders?

c.      What is **post-traumatic stress disorder**?  What are some of the symptoms people experience?

d.      What is a **phobia**?  Do you have any phobias?  If so, what are they and why do you think they occur?  What are some of the various theories on the causes of phobias?

e.      Differentiate between an **obsession** and a **compulsion**.  Describe how a obsessive-compulsive disorder might begin.

f.     What is a somatoform disorder?  Identify and describe the three types of somatoform disorders.

g.     What are some factors that make **conversion disorder** easy to diagnose?

h.     Define dissociative disorder.  Briefly describe and give an example of each of the following: **psychogenic amnesia** and **fugue**.

i.     Give some of the characteristics of **multiple personality disorder**.  What are some of the patterns in the way that multiple personalities develop?

j.      What are some of the thought processes of depressed people?  How do depressed people perceive their environment?

k.      Discuss some of the risk factors involved in depression.

l.      Discuss each of the following theories and explain how each might be related to depression:

   (1)      Psychoanalytic Theory

   (2)      Learning Theory

   (3)      Cognitive Theory

314

(4)    Hopelessness Model

(5)    Medical Model

m.    How do people in a state of **mania** differ from those who are depressed?  Give a brief description of a person with **bipolar disorder**.  What characteristics would they possess?

n.    What disorder plays a major role in **suicide**?  Why?  What are the differences between males and females with regard to suicide?

o.    What is the relation between culture and responses to suicide threats?  How do cultures differ and how are they the same?  What factor does age have?

p.    What are the two main motivations for suicide that have been identified?  What are some of the best predictors of a suicide attempt?

4.    <u>Social</u> <u>Disorders</u>

a.    Why has it been historically difficult for mental health professionals to identify and categorize sexual disorders?  What efforts are being done to remove these issues?

b.    Identify and describe the two general categories that sexual disorders are divided into.  Define the category and list some of its characteristics.

c.    What are some of the social costs of drug abuse?  What do these figures show?

d. How does **substance abuse** result? What makes a person dependent on a substance?

e. What is the result of the misuse of alcohol? What are the two categories of people with regard to alcohol abuse? Describe each one and how they are placed into categories.

f. What are some of the different views as to why some people become addicted to alcohol and develop an uncontrollable urge to drink when others neither become addicted nor feel a need to drink?

g. Identify some of the different types of drugs and give a description of their effect on the body.

h.    What is the history of drug use?  What are some of the medical uses of these drugs?

i.    Contrast physical dependence and psychological dependence.

j.    What is **personality disorder**?  How are they characterized?  Give an original example of an individual with antisocial personality disorder.  What is their personality like and some of their characteristics?

5.    Psychotic Disorders

a.    How do people suffering from **psychosis** react to their environment?

b.      What symptoms must a person show to be diagnosed as schizophrenic?  What is
        the prevalence of schizophrenia in the United States?

c.      Discuss each of the following symptoms of schizophrenia.  Give an example of
        each one.

        (1)     Disorganization of thought

        (2)     Disorganization of perception

        (3)     Disturbances in communication

        (4)     Inappropriate emotions

(5)     Unusual motor activities

d.     Discuss the evidence for a genetic predisposition to schizophrenia.

e.     What has research suggested with regard to the learning approach to schizophrenia?

f.     What is the explanation for schizophrenia from the cognitive perspective?

**POSTTEST**

1. Somatoform disorders involve:
   a. apparent physical illness caused by psychological factors
   b. a tendency to misinterpret minor bodily changes as being indicative of serious illness
   c. genuine physical illness caused in part by psychological factors
   d. the deliberate faking of physical illness

2. What do contemporary practicing clinicians use to diagnose behavior disorders?
   a. *Clinician's Guide to Behavior Disorders*
   b. *Diagnostic and Statistical Manual IV*
   c. *Guide to Psychiatric Diagnosing*
   d. *Handbook of Mental Disorders*

3. The major difference between a phobic disorder and a generalized anxiety disorder is that:
   a. anxiety is specific to one object or situation in a phobic disorder, but is "free-floating" in a generalized anxiety disorder
   b. only the generalized anxiety disorder depends on past conditioning
   c. the generalized anxiety disorder occurs primarily in men and the phobic disorder occurs primarily in women
   d. the phobic disorder is more severe and more difficult to treat

4. Which of the following is classified as a dissociative disorder?
   a. bipolar disorder
   b. fugue
   c. narcissistic personality disorder
   d. obsessive-compulsive disorder

5. The _____ include the attraction to deviant objects or acts that interfere with the capacity for reciprocal and affectionate sexual activity.

6. Four months after being raped, Gloria was found wandering aimlessly in a neighboring town. She was unable to remember her name, address, or family, and she had no memory for the past several days. Gloria was most likely suffering from:
   a. a multiple-personality disorder
   b. a panic disorder
   c. a somatization disorder
   d. psychogenic amnesia

7. A person who maintains bizarre, false beliefs that have no basis in reality is said to have:
   a. delusions
   b. hallucinations
   c. illusions
   d. obsessions

8. With whom is the psychodynamic model of mental illness associated?
   a. Freud
   b. Hippocrates
   c. Kraepelin
   d. Watson

9. The diagnosis of antisocial personality disorder would apply to an individual who:
   a. chronically violates the rights of others
   b. is emotionally cold, suspicious of everyone, and overly concerned about being slighted by others
   c. withdraws from social interaction due to a lack of interest in interpersonal intimacy
   d. withdraws from social interaction due to an intense fear of rejection or criticism

10. Which of the following statements is true with respect to the influence of environment on depression?
    a. failures to achieve goals have their greatest impact on those who are already socially dependent
    b. most depressive episodes are preceded by a stressful life event
    c. problems with other people have their greatest impact on those who are already somewhat self-critical
    d. stressful events can cause depression regardless of the effectiveness of one's coping skills

11. Which of the following individuals would represent the highest risk for attempted suicide?
    a. a 15-year-old male from a wealthy family
    b. a 25-year-old housewife with two children
    c. a divorced, 30-year-old female lawyer with a history of depression
    d. a married, 22-year-old man with a blue-collar job

12. {T  F} Evidence suggests that phobias form best around potential natural threats.

13. Hal thinks constantly about dirt and germs. He washes his hands hundreds of times a day. Hal is most likely suffering from:
    a. hypochondriasis
    b. obsessive-compulsive disorder
    c. phobic disorder
    d. somatization disorder

14. Which model proposes that losing touch with one's feelings, goals, and perceptions forms the basis for psychological disturbance?
    a. cognitive
    b. humanistic
    c. learning
    d. systems

15. What are the most common psychological ailments?
    a. anxiety disorders
    b. mood disorders
    c. personality disorders
    d. schizophrenia

16. The *DSM IV* classification system is said to be "multiaxial." This means that the system:
    a. allows many different potential methods of diagnosing people
    b. asks for judgments about individuals on numerous separate dimensions
    c. is characterized by poor reliability
    d. permits multiple diagnoses of single individual

17. The double-bind conflict as a possible cause of schizophrenia is associated with which of the following models?
    a. cognitive
    b. humanistic
    c. learning
    d. medical

18. Pat is acutely aware of each sniffle he gets -- and is sure each time that it is a sign of cancer or of some other dreaded disease. He runs to the doctor all the time, despite the doctor's assurances that there is nothing seriously wrong with him. Pat has a medicine chest full of medicines. What is the most appropriate diagnosis?
    a. agoraphobia
    b. conversion disorder
    c. dissociative amnesia
    d. hypochondriasis

19. Jack is a top executive in his firm. His workdays are filled with stressful situations. Jack used to have a few drinks socially but now he drinks to deal with stress at work. Jack would be considered a:
    a. compulsive worker
    b. problem drinker
    c. substance abuser
    d. suicidal risk

20. In which perspective is the "symptom" considered to be the sickness?
    a. behavioral
    b. biomedical
    c. psychodynamic
    d. sociological

21. Addiction, including the presence of tolerance or withdrawal, is called

    _____  _____ .

22. The basic problem in the mood disorders is disturbed _____; the basic problem in the schizophrenic disorders is disturbed _____ .
    a. emotion; thought
    b. perception; thought
    c. thought; emotion
    d. thought; perception

23. A given culture's criteria of abnormality:
    a. are affected by current values, trends, politics, and scientific knowledge
    b. are likely to be similar to all other culture's criteria of abnormality
    c. may change with time
    d. both a and b

24. The major difference between a somatization disorder and a conversion disorder is that:
   a. a somatization disorder involves apparent physical illness, and conversion disorder involves genuine physical illness
   b. a somatization disorder involves intentional faking of physical illness, while conversion symptoms are unconsciously created
   c. somatization disorders involve a wide variety of organs and symptoms; conversion disorders involve loss of function in a single organ system
   d. somatization disorders occur only in adults, while conversion disorders occur only in children

25. What is the relative occurrence of depression between the sexes?
   a. it is about equal
   b. it occurs more frequently in women
   c. it occurs much more frequently in men
   d. it occurs slightly more frequently in men

26. The condition in which someone forgets his/her identity and takes on a new identity in a new location is classified as a
   a. catatonic schizophrenia
   b. fugue state
   c. multiple personality
   d. paranoid schizophrenia

27. What is measured by Axis II of the *DSM IV*?
   a. major clinical syndromes
   b. personality disorders
   c. physical disorders relevant to the patient's mental or behavioral problem
   d. psychosocial stressors affecting the patient

28. The finding that many schizophrenics have difficulty in focusing their attention implies that schizophrenia is caused by:
   a. a specific recessive gene
   b. exposure to deviant communication patterns
   c. neurological defects
   d. traumatic childhood experiences

29. People with unipolar disorders experience _____; people with bipolar disorders experience _____.
   a. alternating periods of depression and mania; depression and mania simultaneously
   b. alternating periods of depression and mania; mania only
   c. depression only; alternating periods of depression and mania
   d. mania only; alternating periods of depression and mania

30. Emil Kraeplin's focus on _____ is/are still used in today's classifications of mental disorders.
   a. cultural norms
   b. statistical deviation
   c. symptoms
   d. treatments

31. What is different about the fear response of someone with an anxiety disorder and someone who has a real reason to be afraid?
    a. only a real fear reaction interferes with a person's everyday life
    b. the physiological reactions are very different
    c. there may be no real threat to the person with an anxiety disorder
    d. none of the above

32. The cartoon character Charlie Brown, with his extreme dread, pessimism, worrying, and brooding, could be diagnosed as having a/an _____ disorder.
    a. generalized anxiety
    b. obsessive-compulsive
    c. panic
    d. phobic

33. Most authentic cases of multiple-personality disorder have in common a background of:
    a. a traumatic childhood characterized by physical, emotional, and/or sexual abuse
    b. extremely distorted communication patterns in the family
    c. extremely overprotective parents
    d. having been reinforced for "crazy" behavior

34. Re-experiencing a stressful event that can occur months or years after the event is called _____ stress disorder.

35. Which one of the following is NOT one of the criteria for abnormality?
    a. harmful symptoms
    b. norm violation
    c. strangeness
    d. subjective distress

36. {T F} The inability to consistently experience the normal sexual response cycle is called sexual dysfunction.

37. Hallucinations are a disturbance of:
    a. communication
    b. emotion
    c. motor behavior
    d. perception

38. What distinguishes somatoform disorders from psychophysiological disorders?
    a. in somatoform disorders, there is no identifiable physical problem
    b. in psychophysiological disorders, there is no identifiable physical problem
    c. only somatoform disorders have psychological origins
    d. only psychophysiological disorders have psychological origins

39. Men attempt suicide _____ often than women; men complete _____ suicides than women.
    a. less; more
    b. less; fewer
    c. more; fewer
    d. more; more

40.    Hillary comes from a dysfunctional family. Her mother abuses both her and her brothers. The parents are both substance abusers. Hillary has some very strange behaviors. The assumption that Hillary's problems originated in the family is MOST reflective of the _____ model of mental disorders.
       a.    cognitive
       b.    humanistic
       c.    medical
       d.    systems

41.    On what axis of the *DSM IV* would schizophrenia be placed?
       a.    I
       b.    II
       c.    III
       d.    IV

42.    Which of the following is a mood disorder?
       a.    bipolar disorder
       b.    general anxiety disorder
       c.    phobic disorder
       d.    schizophrenia

43.    Insanity is which type of term?
       a.    legal
       b.    medical
       c.    psychological
       d.    cognitive

44.    {T  F} People with antisocial personality disorder tend to be physically unattractive and socially unskilled.

45.    A person who checks his watch 20 times before going to sleep is being:
       a.    amnesic
       b.    compulsive
       c.    obsessive
       d.    phobic

## POSTTEST ANSWERS

1.      a
2.      b
3.      a
4.      b
5.      paraphilias
6.      d
7.      a
8.      a
9.      a
10.    b
11.    c
12.    T
13.    b
14.    b
15.    b
16.    b
17.    c
18.    c
19.    b
20.    a
21.    substance dependence
22.    a
23.    d
24.    c
25.    b
26.    b
27.    b
28.    c
29.    c
30.    c
31.    c
32.    a
33.    a
34.    post-traumatic
35.    c
36.    T
37.    d
38.    a
39.    a
40.    d
41.    a
42.    a
43.    a
44.    F
45.    b

# PSYCH JOURNAL

**Please use the following pages to record your thoughts and feelings about the following questions.**

1.    The task of identifying and explaining psychological disorders is a complex one. Review the story of Nancy Spungen. Identify those behaviors you find bizarre or "abnormal." As a first step, determine which disorders each of the bizarre behaviors might fall under. You should find that each behavior can be related to several disorders. Next, choose a few of those behaviors and develop explanations for them based on the major approaches (medical, psychoanalytic, learning/behavioral, cognitive, humanistic, systems, and biopsychosocial) to explaining psychological disorders

_____

_____

_____

_____

_____

_____

_____

_____

_____

_____

_____

_____

_____

_____

_____

_____

_____

_____

2. Do you know anyone in your parents' age group who was involved in the Vietnam War? Does that person show any signs of post traumatic stress disorder? What about other people you know who have witnessed a tragedy, natural disaster, act of violence, etc.? Did the event continue to "pop up" in their minds and bother them long after it was over?

_____

_____

_____

_____

_____

_____

_____

_____

_____

_____

_____

_____

_____

_____

_____

_____

3. Have you ever had trouble shaking the blues? How did your "down mood" differ from depression? How were you able to get over it? Can you imagine what it might feel like to always feel down, sad, or hopeless? Has anyone in your family or circle of acquaintances ever suffered from depression? What was it like to be with that person while they were depressed? Did they get help?

_____

_____

_____

_____

_____

_____

_____

_____

_____

_____

_____

_____

_____

_____

_____

_____

_____

_____

4. Alcohol use seems to be everywhere on most college campuses today. Do you think the alcohol industry glamorizes drinking on college campuses? Do you know anyone with a serious drinking problem now? How do you differentiate casual alcohol use from alcoholism?

_____

_____

_____

_____

_____

_____

_____

_____

_____

_____

_____

_____

_____

_____

_____

_____

_____

_____

_____

_____

_____

5. Does alcoholism or drug addiction run in your family? Among the people who know, what kinds of life problems have been caused by excessive drug or alcohol use?

_____

_____

_____

_____

_____

_____

_____

_____

_____

_____

_____

_____

_____

_____

_____

_____

_____

_____

_____

# Chapter 14 -- Therapy

**CHAPTER SUMMARY**

1.      Two broad classifications of therapy deal with psychological disturbances: Psychotherapy is the use of psychological techniques by a professionally trained individual to help a client change unwanted behavior and adjust to his or her environment. Biomedical therapy, which may be used together with psychotherapy, includes the use of drugs, surgery, and electric shock to induce behavior change. Historically, many treatment methods have been quite inhumane. Reforms of asylums began in the late 1700s. The World War II era led to the use of group therapy and tranquilizers. Today, deinstitutionalization and prevention are major mental health policies. Therapies can be divided into two categories. Insight therapies are aimed towards helping individuals become aware of the motives, often repressed or denied, that are shaping their behaviors. Action therapies do not focus on digging deeply into the individual's personality. Action-oriented therapists direct their attention to the presenting symptoms and attempt to change the specific problematic behaviors or beliefs.

2.      Psychoanalysis, developed by Sigmund Freud, is based on the belief that events in early childhood are important determinants of one's personality. Feelings or events associated with early childhood may be repressed, leading to anxiety later on. These repressed feelings enter the person's unconscious and can be examined through the use of free association or dream interpretation. Carl Rogers and other humanistic psychologists focus on helping people through a process of self-growth. Through client-centered therapy, the therapist uses unconditional self-regard to improve the client's functioning and to increase the client's self-awareness, self-reliance, and comfort with relationships. Gestalt therapy proposes that unconscious thoughts and emotions (background) may lead to behaviors (figure) that are inappropriate. The Gestalt therapist focuses on the present and uses a number of techniques including group settings and role playing to facilitate self-awareness.

3.      Behavior therapies assume that abnormal behaviors are learned; thus, they can be unlearned by changing the events that reinforce maladaptive behaviors. Behavior therapists may use a variety of techniques to change behavior. Aversive conditioning uses classical conditioning to stop undesirable behaviors. Based on the assumption that it is impossible to be relaxed and anxious at the same time, systematic desensitization involves making the client relax in the presence of an object or event that used to cause anxiety. With operant conditioning, positive rewards are used to motivate people to perform desirable behaviors. Another method of changing behaviors involves modeling. With this type of behavior therapy, the client watches someone else engage in the desired behavior and imitates them. According to rational emotive therapy (RET), a cognitive therapy, our problems are not the result of how we feel; rather, how we think and believe determines how well we will adjust to our environment. Focusing on the present, the RET therapist attempts to help the client recognize the illogic of his or her thoughts, and thus change them.

4.      Group therapy has many advantages. It is more economical than individual therapy. Since many clients' problems involve interpersonal relationships, the group setting is the best place to examine these problems and "practice" interpersonal behaviors. Group members encourage one another and show each other that they are not alone in having problems. Group members can help identify each

other's problems and offer solutions. Group therapy also has disadvantages. Some people may need individual therapy before they are able to function in a group. Confidentiality may be an issue as well. In having to deal with a group, the therapist may not be giving each client enough attention. Finally, the pressure to conform to group rules may limit the therapy process.

5. Electroconvulsive therapy (electric shock treatment) is given most often to people who are suffering from severe depression, have not responded to drug treatment, and who may attempt suicide if not relieved of their depression. Psychosurgery involves cutting connections between different parts of the brain; the prefrontal lobotomy is the best-known type. Most mental health practitioners view psychosurgery as barbaric. Clinical psychopharmacology is the use of drugs (antipsychotic, antidepressant, antianxiety) to treat psychological disorders. Although these drugs are being used successfully in treating mental illness, they all have undesirable side effects; clients must be constantly monitored.

6. Therapy can be very useful for treating many psychological disorders, but everyone will not be helped by therapy. The factors that determine therapy effectiveness are characteristics of the client, characteristics of the therapist, the nature of the problem, and the type of therapy. Some therapies are useful for a very specific type of client whereas others have broader application. Consequently, many therapists adopt the eclectic approach which bases the choice of therapy technique on the demands of the individual client.

7. Community psychology is based on the prevention of mental health problems and treating people in community-based programs. Prevention includes three levels: Primary prevention is aimed towards preventing the development of mental health problems. Secondary prevention aims to identify and treat problems early in order to reduce their severity. Tertiary prevention involves the rehabilitation and follow-up treatment of people who have been hospitalized or have been in long-term treatment. Deinstitutionalization has been linked to the growing problem of the homeless. Many of the people released from hospitals are ill-equipped to function successfully in the community.

## KEY TERMS AND CONCEPTS

The History and Scope of Therapy
    Two broad classifications of therapy that deal with psychological disturbances
        Psychotherapy - Use of psychological techniques to help a client
        Biomedical therapy - Use of drugs, surgery, or electric shock
    Goals of therapy
        Techniques are safe and do not cause more problems
        Reduce individual's present discomfort
        Aid in the development of a healthier individual
    Historically, many treatment methods have been inhumane
        Reforms of asylums began in late 1700s
        World War II era led to the use of group therapy and tranquilizers
        Today, deinstitutionalization and prevention are utilized
    Approaches to psychotherapy
        Insight therapies
            Help individuals become aware of motives shaping behaviors
            Psychoanalysis, client-centered therapy, and gestalt therapy

Action therapies
    Focus on presenting symptoms and change behaviors

Insight Therapies
    Psychoanalysis
        Events early in life are important to one's personality
        Repressed emotions are examined through free association and dreams
        Transference - transfer to therapist feelings originally aimed at others
    Humanistic therapy
        Carl Rogers and others used the process of self-growth
            Therapist does this by providing a positive setting
                Give clients unconditional positive regard
                Therapist must be genuine or real
                Therapist shows empathy
        Use client-centered therapy - focus on client rather than therapist
            Goals of client-centered therapy
                Increase self-awareness
                Increase self-acceptance
                Increase comfort with relationships
                Increase self-reliance
                Improve functioning
    Gestalt therapy
        Awareness is the focus
        Unconscious thoughts and emotions may lead to inappropriate behaviors
        Utilize group settings and role playing to help self-awareness

Action Therapies
    Behavior therapies - abnormal behaviors are learned and thus can be unlearned
        Aversive conditioning
            Classical conditioning used as its base for changing behavior
            Used to stop undesirable habits such as smoking and drinking
        Systematic desensitization - treats phobias and anxiety disorders
            Client is told to relax in the presence of an object causing anxiety
            Steps of systematic desensitization
                Events or objects that causes stress are identified
                Teaching deep muscle relaxation
                Pairing of relaxation with the stressful events or objects
        Operant conditioning therapies
            Positive rewards are used to motivate desirable behaviors
            Therapy using these procedures is behavior modification
                Token economies are used in this type of therapy
        Modeling therapies
            Client watches someone else engage in desired behavior
    Cognitive therapies: an emphasis on thought, beliefs, and attitudes
        Changes in thought will lead to changes in behavior and feelings
        Rational emotive therapy (RET)
            Helps client recognize the illogic of his/her thoughts
            Various techniques - referenting, semantics, humor, etc.

Psychotherapy in Groups
    Advantages of group psychotherapy
        More economical than individual therapy
        Group members encourage one another and show that they are not alone
        Group members can help identify each other's problems and offer help

Disadvantages of group therapy
    Some may need the individual therapy
    Confidentiality is an issue
    Pressure to conform to group rules may limit therapy process
Family therapy - working with the family as a unit (parents and children)

Biomedical Therapies - assumes many disorders are result of biological components
    Electroconvulsive therapy
        Some of electric shocks uses
            Those with severe depression
            Those who haven't responded to drug treatment
    Psychosurgery - Cutting connections between different parts of the brain
        Prefrontal lobotomy is the best-known type
        Most view psychosurgery as barbaric
    Clinical psychopharmacology - use of drugs to treat psychological disorders
        Undesirable side effects may result with use of these drugs
        Psychotropic drugs - chemicals used to treat disorders
            Treatment of psychoses
            Treatment of depression
            Treatment of mood disorders
            Treatment of anxiety

Does Therapy Work?
    Not everyone can be helped by therapy
    Factors that determine therapy effectiveness
        Characteristics of the client
        Characteristics of the therapist
        Nature of the problem and type of therapy
    Eclectic approach - used to avoid becoming locked into one type of therapy
    Multicultural diversity and therapy effectiveness
        Different characteristic exists for various groups
            African americans
            Asian americans
            Hispanic americans
            Native americans

Community Psychology - based on prevention and treating in community settings
    Primary prevention - preventing the development of disorders
    Secondary prevention - identify and treat disorders early
    Tertiary prevention - rehabilitation and follow-up treatment

## DISCUSSION QUESTIONS AND EXERCISES

**Note: These questions and exercises ARE the learning objectives for this chapter. Answer them accurately in your own words and you will have mastered the most important material. We guarantee it.**

1.   The History and Scope of Therapy

a.   In your own words provide a description of both classifications of therapy that are aimed at dealing with psychological disturbances.

b.   What are the three basic goals of therapy?

c.   Discuss the history of therapy. How did it develop? How has it changed? What practices are used today?

d.    Contrast insight therapies with action therapies.

2.    <u>Insight Therapies</u>

a.    What is the psychoanalytic view on how psychological disorders occur?  What are the two methods Freud recommends to release repressed experiences?

b.    Define **transference**.  Give your own example of this phenomenon.  Why is it an important part of therapy?

c.    Many humanistic psychologists believe that the root of many disorders is in childhood.  However, their approach is very different.  How is the humanist approach different from psychoanalysis?  What is the goal of therapy?

d.   In order to achieve a setting in which people do not fear social rejection or expressing themselves, what three conditions are necessary?

e.   In humanist therapy the emphasis is on the client rather than on the therapist. What is this called and give an example of this approach?

f.   What are the goals of client-centered therapy?

g.   How does Gestalt therapy propose we improve our adjustment? What is the central focus of Gestalt therapy? Why?

3. <u>Action</u> <u>Therapies</u>

a. When did behavior therapies become prominent? What happened during this time period?

b. What is aversive conditioning? How has it been used? Give an example of a undesirable behavior and an aversive method to break it.

c. A number of techniques have been developed to treat phobias and other anxiety disorders. Describe **systematic desensitization**. What are the three steps?

d.    The use of operant conditioning procedures in therapy is referred to as **behavior modification**.  Give your own example of a **token economy**, go through all the steps.

e.    What are some of the advantages of behavior modification?  What are some of the criticisms made by those who favor other approaches?

f.    Albert Bandura used a technique to reduce a wide range of phobic behaviors. What is this technique?  How can it be used and give an example?

g.    What is the aim of cognitive therapy?  Discuss **rational emotive therapy (RET)**.  What is it and how is it used?

h.     Name and describe at least five of the techniques used by rational emotive therapists.

4.     <u>Psychotherapy</u> <u>in</u> <u>Groups</u>

a.     Although the aim of both individual and group therapy is the same, there are some important differences between the two settings.  What are they?

b.     Identify some of the advantages and disadvantages of group therapy.  What is **family therapy?**

5.     <u>Biomedical</u> <u>Therapies</u>

a.     What is the basis of biomedical therapies?  How are they utilized?

b.     Describe the process of **electroconvulsive therapy**.  Does it work?  Clearly there are some pros and cons to ECT, but what is the American Psychiatric Association's position on ECT?

c.     What is **psychosurgery**?  What is the surgical technique developed to control the emotions experienced by those with psychological disturbances?

d.     What has research on lobotomies found?  Is it still used today?  What are your thoughts on psychosurgery?

e.      What is clinical psychopharmacology?

f.      Give a brief description of each of the following treatments done in clinical psychopharmacology:

   (1)      Treatment of psychoses

   (2)      Treatment of depression

   (3)      Treatment of mood disorders

   (4)      Treatment of anxiety

6.      <u>Does Therapy Work</u>

a.      What are the four factors suggested for a successful treatment?  Are these factors good enough to predict the failure or success of therapy?

b.      What is the **eclectic approach**?

c.      There is a growing concern that the mental health system is not fully responsive to the mental health need of cultural minorities. Characterize each of the following cultural groups and discuss what a therapist might do to help an individual from each group.

        (1)      African Americans

        (2)      Asian Americans

        (3)      Hispanic Americans

        (4)      Native Americans

7.      <u>Community Psychology</u>

a.      What are the cornerstones of **community psychology**?

b.  Discuss each of the following levels of prevention as defined by community psychology.

   (1)  Primary prevention

   (2)  Secondary prevention

   (3)  Tertiary prevention

c.  What are some of the criticisms of this type of therapy?  What policy has been linked to the growing problem of the homeless?  Why?

## POSTTEST

1.  With what therapeutic approach is dream analysis most closely associated?
   a.  behavior modification
   b.  cognitive
   c.  humanistic
   d.  psychoanalytic

2.  One of the most important concerns in group therapy is:
   a.  cost
   b.  confidentiality
   c.  educational level of the group
   d.  the number of individuals in the group

3. Psychopharmacology involves the treatment of mental disorders with:
   a. electroconvulsive shock
   b. insight therapy
   c. medication
   d. systematic desensitization

4. On average, what are the results of outcome studies on the effectiveness of psychotherapy?
   a. moderately negative
   b. moderately positive
   c. very negative
   d. very positive

5. Which of the following is NOT one of the three goals of all therapies?
   a. curing the patient
   b. decreasing client discomfort
   c. improving the client's adjustment
   d. insuring that therapy does not harm the client

6. {T F} Family therapists view abnormal behavior as simply maladaptive and serving no purpose.

7. Homes where people live after they have been discharged from an institution are called _____ _____.

8. Which of the following statements is most accurate?
   a. Although many people have benefited from deinstitutionalization by avoiding unnecessary hospital stays, there have been some unanticipated problems
   b. Deinstitutionalization has been a failure, resulting in poorer treatment for both inpatients and outpatients
   c. Deinstitutionalization has been an outstanding success, resulting in better care for all clients.
   d. Deinstitutionalization hasn't had much of an impact on mental health care at all.

9. Drug therapy is classified as a(n) _____ therapy; electroconvulsive shock therapy is classified as a(n) _____ therapy.
   a. aversion; biomedical
   b. behavior; insight
   c. biomedical; behavior
   d. biomedical; biomedical

10. Electro-convulsive shock therapy is currently used predominantly for the treatment of:
   a. amnesia
   b. multiple personalities
   c. schizophrenia
   d. severe depression

11. Which of the following statements about "self-help" books is TRUE, according to the text's authors?
   a. Caution should be used in trying self-help advice.
   b. They are good for diagnosing problems.
   c. They are good for identifying problems.
   d. They cannot make an existing problem any worse.

12. The evidence suggests that some psychotherapies are better for some problems than others. For example, behavior therapies such as systematic desensitization works best for phobias. What conclusions can be drawn from this research?
   a. an eclectic approach to therapy might be in the therapist's and client's best interest
   b. psychotherapy is almost impossible to seek because one is never sure which therapist is best for one's own particular problem
   c. psychotherapy is impossible to employ because one is never sure what technique is best
   d. none of the above

13. The therapeutic technique in which the patient says whatever comes into his or her mind is:
   a. free association
   b. freedom of speech
   c. systematic desensitization
   d. the search for the self

14. Operant conditioning can be used in classrooms, hospitals, prisons, and industry by giving individuals exchangeable rewards for performing appropriate actions. This illustrates which of the following?
   a. bartering
   b. bribery
   c. negative reinforcement
   d. token economy

15. Antianxiety drugs:
   a. are prescribed only for people who have a clinical anxiety disorder
   b. can permanently cure anxiety disorders
   c. can temporarily alleviate feelings of anxiety
   d. help control bodily functions

16. Which therapeutic approach is characterized by the view that it is unreasonable "shoulds, oughts, and musts" that people have as internal standards that cause them to develop psychopathy?
   a. covert sensitization
   b. client-centered therapy
   c. rational-emotive therapy
   d. systematic desensitization

17. _____ therapy helps people to control their physiological functions without the use of drugs.

18.  What is the basic premise of behavior modification therapy
     a.    psychopathology arises because, instead of living up to their own
           self-perceptions, people try to live up to what others want them to be
     b.    psychopathology arises because of the way people evaluate what happens
           to them
     c.    psychopathology is a reflection of unconscious conflicts, and these
           conflicts must be made conscious so they can be better handled
     d.    psychopathology is learned and therefore can be unlearned

19.  Rogers believed that client-centered therapists must provide unconditional
     positive regard for their clients.  In other words, they must:
     a.    communicate with clients in an honest and spontaneous manner
     b.    provide warmth and caring only when clients' behavior is appropriate
     c.    show complete, nonjudgmental acceptance of the client as a person
     d.    understand the client's world from the client's point of view, and be able
           to communicate this

20.  Which of the following is a component of community psychology?
     a.    biofeedback
     b.    emphasis on prevention
     c.    free association
     d.    the search for the true self

21.  Early intervention is referred to as _____ in community psychology
     model.
     a.    primary prevention
     b.    secondary prevention
     c.    tertiary prevention
     d.    outpatient psychotherapy

22.  For which of the following behaviors would aversive conditioning therapy be
     best?
     a.    alcoholism
     b.    forgetfulness
     c.    phobic avoidance
     d.    shyness

23.  The medically based therapy/therapies most commonly used today is/are
     a.    cingulotomy
     b.    electro-convulsive shock therapy
     c.    psychoactive drugs
     d.    psychosurgery

24.  Gestalt therapists encourage clients to put emphasis on:
     a.    changing the client's beliefs
     b.    changing the client's value system
     c.    the collective "we"
     d.    the first-person pronoun

25.  The risks of ECT:
     a.    are negligible, as long as appropriate precautions are taken
     b.    are so severe that the use of ECT has been banned by law
     c.    are completely eliminated by modern improvements in the procedure
     d.    include both short- and long-term intellectual impairment

26. Which of the following is NOT among the advantages of group therapy?
    a. Certain kinds of problems are especially well-suited to group treatment
    b. It produces a significantly higher recovery rate than individual therapy
    c. It provides an opportunity for participants to work on social skills in a safe environment
    d. Participants often come to realize that their misery is not unique

27. The main component of behavioral approaches to therapy is:
    a. conversations with the therapist in which the patient tries to recall and reexperience unpleasant memories
    b. structured exercises to overcome the person's embarrassment about openly discussing feelings
    c. tests designed to discover what the patient is really like
    d. the use of classical or operant conditioning techniques to replace abnormal or maladaptive behavior with socially accepted behavior

28. In the past 25 years and continuing today, the trend in treatment of mental disorders is toward:
    a. cures through the use of intensive psychoanalysis
    b. deinstitutionalization
    c. large, well-equipped mental hospitals
    d. more widespread use of psychosurgery

29. What perspective on abnormal behavior stresses that it is not the actual events that happen to people that make them upset, but the way people view those events?
    a. behavior modification
    b. cognitive
    c. community
    d. psychoanalytic

30. Whose therapy is sometimes called "client-centered"?
    a. Bandura
    b. Freud
    c. Rogers
    d. Watson

31. With what therapeutic approach is an attempt to help the client get in touch with his or her basically good inner core most closely associated?
    a. cognitive
    b. humanistic
    c. psychoanalytic
    d. social learning

32. Bandura and his associates have used modeling techniques to:
    a. bring repressed memories to the surface
    b. reduce phobic behaviors
    c. reduce symptoms of schizophrenia
    d. treat generalized anxiety

33. {T F} One of the major problems with the antianxiety drugs is that they are addicting.

350

34. What is the basic premise of psychoanalytic therapy?
   a. psychopathology arises because, instead of living up to their own self-perceptions, people try to live up to what others want them to be
   b. psychopathology arises because of the way people evaluate what happens to them
   c. psychopathology is a reflection of unconscious conflicts, and these conflicts must be made conscious so they can be better handled
   d. psychopathology is learned and therefore can be unlearned

35. Compared to Freudian psychoanalysis, modern psychodynamic therapy:
   a. does not utilize free association or dream analysis
   b. is usually longer term
   c. usually involves more direct, face-to-face interaction
   d. usually puts more emphasis on early childhood experiences

36. With what specific therapeutic technique is the construction of anxiety hierarchies and the teaching of deep muscle relaxation associated?
   a. aversive conditioning
   b. free association
   c. one-on-one interaction
   d. systematic desensitization

37. In the Middle Ages, it was believed that people with emotional problems:
   a. had low intelligence
   b. were "chosen" by a higher authority
   c. were contaminated by exotic germs
   d. were possessed by demons

38. Antipsychotic drugs:
   a. are effective in about 95% of psychotic patients
   b. are often prescribed even for individuals who have no clinical psychotic disorder
   c. gradually reduce psychotic symptoms such as hallucinations and delusions
   d. tend to produce an immediate, but short-term effect

39. Who considered dreams the road to the unconscious?
   a. Adler
   b. Ellis
   c. Freud
   d. Rogers

40. In community psychology, _____ is designed to identify and change the sources of mental health problems.
   a. primary prevention
   b. secondary prevention
   c. tertiary prevention
   d. treatment prevention

41. When patients treat the therapist as the symbolic equivalent of important figures in their lives, it is called:
   a. primary transference
   b. resistance
   c. systematic desensitization
   d. transference

42. Behavioral therapies based on operant conditioning:
   a. associate problem behavior with negative or positive experiences
   b. challenge the client to change
   c. use rewards or punishments to bring about change
   d. use unconditional positive regard

43. When treating a patient, the patient has problems when he or she gets upset every time someone shows some displeasure with him or her. A humanist would do which of the following:
   a. delve into the patient's childhood to find out about the patient's relationship with his or her parents
   b. point out to the patient that it is irrational to expect people not to show displeasure with others on occasion, and when it does happen it isn't so bad
   c. reflect back to the patient the real depth of the anger and pain that the therapist believes the patient really feels but is not acknowledging
   d. teach the patient a snappy "comeback" to use with such people

44. Client "I've had a bad week. I'm really unhappy." Therapist: "You've had some unpleasant experiences this week and are feeling quite depressed as a result." The therapist's statement in this interaction is intended to communicate _____ to the client.
   a. disapproval
   b. empathy
   c. genuineness
   d. unconditional positive regard

45. The main component of insight therapy is:
   a. conversations with the therapist in which the patient tries to recall and re-experience unpleasant thoughts
   b. structured exercises to overcome the person's embarrassment about openly discussing feelings
   c. tests designed to discover what the patient is really like
   d. the use of classical or operant conditioning techniques

# POSTTEST ANSWERS

1. d
2. b
3. c
4. b
5. a
6. F
7. Halfway houses
8. a
9. d
10. d
11. a
12. a
13. a
14. d
15. c
16. c
17. Biofeedback
18. d
19. c
20. b
21. b
22. a
23. c
24. d
25. d
26. b
27. d
28. b
29. b
30. c
31. b
32. b
33. T
34. c
35. c
36. d
37. d
38. c
39. c
40. a
41. d
42. c
43. c
44. b
45. a

## PSYCH JOURNAL

**Please use the following pages to record your thoughts and feelings about the following questions.**

1. Review the psychotherapy used to treat Chris Sizemore. Now, using the discussion of the various types of psychotherapies, envision how the treatment program would have proceeded using each of these methods. Then outline an eclectic approach that would combine elements from many of the different approaches.

_____

_____

_____

_____

_____

_____

_____

_____

_____

_____

_____

_____

_____

_____

_____

_____

_____

_____

_____

_____

2.	Have you ever trained yourself to break a bad habit like biting your nails or smoking?  How did you go about breaking yourself of unwanted behaviors?

_____

_____

_____

_____

_____

_____

_____

_____

_____

_____

3.	Have you ever talked a friend (or yourself) through a tough situation or disappointment in order to realistically asses the event and reduce or process the emotional punch they were feeling?  Describe such a conversation.

_____

_____

_____

_____

_____

_____

_____

_____

_____

4. Do you have a group of friends with whom you can discuss problems that you all experience? What advantages and disadvantages do you see with sharing your outlook and problems with a group of people.

_____

_____

_____

_____

_____

_____

_____

_____

_____

5. Has any member of your family (or close circle of friends) ever gone through therapy? Did your whole family shift or change as a result of that one person's therapeutic growth?

_____

_____

_____

_____

_____

_____

_____

_____

_____

_____

_____

6.     Do you know anyone who takes psychotropic drugs? For what kind of disorder? Does that person also participate in therapy of some kind? If you knew the person before they started taking medication, do you see any change in the person's behavior or outlook since they have been taking the drugs?

_____

_____

_____

_____

_____

_____

_____

_____

_____

_____

_____

_____

_____

_____

_____

_____

_____

_____

_____

7. What therapeutic options are available to you around campus, should you face personal challenges that you feel are too much to shoulder alone?

_____

_____

_____

_____

_____

_____

_____

_____

_____

_____

_____

_____

_____

_____

_____

_____

_____

_____

_____

_____

# Chapter 15 -- Interpersonal Relations

## CHAPTER SUMMARY

1.      Social psychology as we know it today began in the 1930s when Kurt Lewin conducted several experiments showing how people were influenced by others. In the mid 1950s Leon Festinger introduced the theory of cognitive dissonance. In the 1970s the field expanded to explore areas such as interpersonal relations (attraction, aggression, and love), group dynamics, intergroup relations, helping, and impression formation. Today's emphasis on cognitive social psychology examines how people process information and how this processing affects their behavior in social situations.

2.      According to the social cognition model, forming impressions involves processing information based on physical appearance, behavior, traits, and characteristics. We are not passive consumers of information about others. We seek and organize that information based on our expectations and schemas. As a result, we often make errors in the way we characterize others in our minds. The attribution process involves determining what a behavior reveals to us about another person. In making attributions we must determine whether a behavior is caused by the situation or the person. There are many biases that affect the attribution process. These biases include a greater likelihood of assigning a trait when we are affected by the behavior (underemphasizing the role of the situation), and the tendency to attribute another's behaviors to traits while viewing our own actions as a response to the situation. The first information we receive has greater influence than later information on our impressions. This is called the primacy effect. The self-fulfilling prophecy refers to the fact that we often unwittingly act to bring about the situations we expect. We then use the results to reinforce our expectations. Attributions about ourselves are also biased. We tend to exaggerate the importance of our contributions in shaping events.

3.      Attitudes are relatively enduring feelings about objects, events, or issues. They often include evaluations, beliefs, and a behavioral component. Parents have the strongest influence on many of our attitudes, because they are automatically believed and they do not have to deal with preexisting attitudes. Personal experiences, culture, peers, the media, and even genetics also influence our attitudes. The Elaboration Likelihood Model (ELM) suggests that attitudes are changed through two routes. The central route involves scrutinizing and evaluating the message content. The peripheral route involves issues outside the message content, such as the characteristics of the communicator and our mood. Attitudes that are salient and those that are very specific are more likely to influence behavior. Furthermore, attitudes that are based on personal experience are the best predictors of behavior.

4.      Cognitive dissonance theory is based on the assumption that people strive to have their attitudes, beliefs, and behaviors support one another. When these elements are not in accord, people often change their cognitions or attitudes to be consistent with their actions.

5.      We enjoy relationships with people who reward us and to whom we also give rewards. Similarity is one of the strongest predictors of attraction because similar others validate our own opinions and actions. Also we expect to have a positive interaction with similar others, a lot in common to share. However, we are not attracted to people who have a characteristic that we do not admire in ourselves. We like people who are physically close to us, although proximity can also lead to

disliking. We like physically attractive people because they may give us a certain status and we see ourselves as being more similar to attractive people than to unattractive people. Also attractive people often have more pleasing personalities and a higher sense of well-being than unattractive people. However, we often end up with partners who are about as attractive as we are, because we like people who are attracted to us. Friendships go through several stages including proximity, similarity, rewardingness, development of regular meetings, self-disclosure and, sometimes, decline. Culture and gender influence how we act toward our friends.

6.      Helping occurs from one of two motivations: altruism or egoism. Helping is influenced by rewards, mood and empathy, and norms. Research on bystander intervention shows that the decision to help in emergency situations is the need result in a complex chain of events. Before helping will occur, people need to notice the person in distress, interpret the situation as an emergency, assume responsibility to help, know the appropriate way to help, and decide to offer aid. A negative decision at any stage assures no helping. An individual is less likely to help if there are others present or if he or she believes the victim does not deserve to be helped.

## KEY TERMS AND CONCEPTS

Scope and History of Social Psychology
    Social psychology focus is on the common ground between people
    Field of social psychology
        In the U.S. only at first, but has grown elsewhere
        Lewin - first social psychology experiment
        Festinger's cognitive dissonance theory was presented

Perceiving and Evaluating Others
    Gathering and organizing social information
        Social cognition model (how we form impressions)
            Physical appearance
            Behaviors
            People's traits
        Schemas - an organized body of knowledge about people and events
            Offers us a short-cut to store information
            Common schema is the stereotype
    Attribution process - interpreting the information
        Situations versus personal traits
            What causes people's actions - disposition or situation
            We base our decision on three elements
                Distinctiveness, how others respond, and consistency
        Assigning traits to individuals
            Once the situations is ruled out, we go to the next step
                Based on information from three areas we assign
                     Knowledge, ability, and intention
    Tendencies and biases in attribution
        Biases affect how we act and how we make attributions
        Hedonic relevance
            Degree in which a person's behavior is costly or rewarding
            Influences attributions in two ways
                Increases likelihood that a person's behavior is a trait
                Increases the extremity of the evaluation

Overemphasizing behavior/Underemphasizing context
    Fundamental attribution error
        Basing attributions solely on behavior
        Occurs in a wide variety of situations
    Looking out or looking in: Actor-observer differences
        Other behaviors caused by internal traits
        Our own behavior caused by situation we are in
Order effects
    Primacy effect
        First information has greater influence than later information
        Strongest when we are judging stable characteristics
Getting what we expect: self-fulfilling prophecy
    Expectation that leads one to behave in ways to cause the expectation
Knowing ourselves
    Attributions about ourselves are often biased (egocentricity)


Our Beliefs and Attitudes
    Attitudes - relatively enduring feelings about objects, events, or issues
    Three components of attitudes
        Evaluation - positive or negative meaning on the object or event
        Belief - statements that express a relationship between events
        Action - how people should act toward an object
    Development of attitudes
        Parents - influence attitudes with reward and punishment
            Powerful sources of control
            Control the information that reaches their children
            This is a strong and lasting influence
        Peers - supply child with new information and different views
            Also utilize the threat of rejection (act and believe as I do)
            This is true of adults also
        Personal experience and culture
            Our strongest and most difficult to change come from experience
            View our own attitudes in light of our culture also
        The media
            Able to reach many people
            Often the only source of information we have about events
        Genetics
            These are our strongest and most stable attitudes
    Attitude change - accepting a new position while giving up an old one
        Elaboration Likelihood Model - two routes to persuasion
            Central route - content of the message
            Peripheral route - outside the content of the message
        Communicator - characteristics are very important
            Factors that determine the effect of the communicator
                Credibility - believability
                Trustworthy
                Similarity
        Message - factors about the message that help
        Audience - few characteristics have been identified to help
    Relationship between attitudes and behavior
        Attitudes do not always predict behavior
        Attitudes central in attention influence behavior more
        The more specific the attitude the more likely it will guide our behavior

When Behaving is Believing
    Theory of cognitive dissonance - Festinger's theory
        State that exists when attitudes and behaviors are in conflict
        To relieve this, a person may change cognitions so they agree
            Effort justification - suffering leads to liking
            Just-barely-sufficient threat
                We derogate objects without strong external constraints

Liking: Positive Attitudes about People
    Attraction - a positive attitude about someone else
    Rewards: receiving and giving
        Most satisfying relationships are those that offer receiving and giving
        Two general types of relationships
            Exchange relationships - expect that rewards will be reciprocated
            Communal - meeting needs of the other person
                Males may be more exchange oriented
                Culture may also be an influence
    Similarity: birds of a feather
        We are attracted to those who are similar
        We are repelled by those who we have little in common
            Similar others validate our own opinions and actions
            We expect to have a positive interaction with similar others
            We expect similar others to like us
    Proximity: To be near you
        Those who are close are more likely to reward each other
        More comfortable to like these people than to dislike them
    Physical attractiveness: the way you look
        We like attractive people for the status they can bring to us
        We see ourselves as being more similar to attractive people
        Attractive people have more pleasing personalities
    Tides of friendship
        The first phase of friendship is proximity
        Friendship flourishes as we find grounds of common interest
        The parties find that they have something to offer each other
        Developing regular meetings (dining, sports, etc.)
        Self-disclosure develops with trust
        Decline occurs in many cases of friendships
            Factors that influence the friendship process
                Gender and culture

Helping Behavior
    Why do people help...or not help
        Helping results from altruism and egoism
        Factors that influence when people help
            Rewards - cost is low and reward is high
            Mood and empathy - good moods lead to helping
            Norms - general rules of behavior that apply to everyone
    Decisions to help (bystander intervention) - a decision tree
        Noticing the victim
        Interpreting the situation as an emergency
        Assuming responsibility
        Knowing how to help
        Taking action

# DISCUSSION QUESTIONS AND EXERCISES

**Note: These questions and exercises ARE the learning objectives for this chapter. Answer them accurately in your own words and you will have mastered the most important material. We guarantee it.**

1.    The History and Scope of Social Psychology

a.    What is **social psychology** and what is its focus?

b.    Social psychology is a relatively young branch of psychology. Discuss the history of social psychology from when it began through its emphasis today.

2.    Perceiving and Evaluating Others

a.    Describe the process of gathering and storing information. Provide each of the three steps and describe how they work together.

b.    Because we see hundreds of people and experience many things, we develop a convenient way to store all of this information. Identify the system we use for storing all this information and describe how we use it.

c.    What are stereotypes? Give an example of a stereotype and identify some of the types of errors it may represent.

d.    Define **attribution**. What is the relationship between the situation a person is involved with and his or her personal traits?

e.    We base our decision on what caused a person's behavior on three elements: distinctiveness, consensus, and consistency. Discuss each one of these elements and give an original example of each element.

f.    How do we move from observing behaviors to assigning specific traits to people?  Identify the steps in this process.

g.    What is **hedonic relevance**?  In what way does it influence attributions?

h.    Define **fundamental attribution error.**  When are we likely to make this error?  Why do we ignore situational factors?

i.    What reasons have been suggested to explain why there is a general tendency for people to view the behavior of others as being caused by internal traits, while we see our own behavior as being determined by the situation we are in?

j.  What is the **primacy effect**?  When are primacy effects the strongest?  Provide some of the explanations for the primacy effect.

k.  What evidence does research provide about our expectation about people and events and the eventual fulfillment of these expectations?

l.  How do we determine our own attitudes?  Are our attributions about ourselves biased?  If so, why?  If not, why?

3.  Our Beliefs and Attitudes

a.  How do most investigators view **attitudes**?  What are the three components of attitudes?

b.    Because our attitudes are influenced by multiple sources, we often hold attitudes that are different from each other.  For each of the following sources, describe how each one affects our attitudes:

(1)    Parents


(2)    Peers


(3)    Personal experience and culture


(4)    The media


(5)    Genetics



c.    As we get older, we are no longer dealing with the formation of new attitudes but rather with the effort to bring about an **attitude change**.  Identify and describe the model which argues that there are two routes to persuasion.

d.    Identify and describe each of the three issues that occur when we examine attitude change.  How does each affect us and why?

e.    What is the relationship between our attitudes and our behavior?

4.    <u>When</u> <u>Behaving</u> <u>is</u> <u>Believing</u>

a.    Discuss **cognitive dissonance**.  What is it, and how is it relieved?  When is dissonance most likely to occur?

b.    Identify and describe two predictions that are based on the theory of cognitive dissonance.

5. Liking: Positive Attitudes about People

a. What role do rewards play in explaining why people are attracted to each other? Describe the two general types of relationships that are distinguished by the expectations of reward exchange.

b. How do gender and culture affect rewards and giving?

c. What is the strongest predictor of attraction? Why? Identify some of the various reasons for the similarity-attraction effect?

d.      Why does proximity lead to liking?  Does proximity always lead to liking?

e.      Why does physical attractiveness have such a strong effect?  Does this mean that unattractive people are doomed?

f.      Discuss the different phases of friendship.  How do they begin and eventually how might they end?

g.      What factors might influence friendship?  Why?

6.   Helping Behavior

a.   What are the two motivations that result in helping behavior?  Provide an original example of these two concepts.

b.   Identify and give a brief description of each of the factors that have been shown to influence when people help.

   (1)   Rewards

   (2)   Mood and empathy

   (3)   Norms

c.       What is **bystander intervention**?

d.       Identify and discuss each of the five steps that are involved in the process of helping (or not helping).

## POSTTEST

1.       In Kelley's attributional model, the dimension of consistency refers to whether:
   a.       a person's behavior is unique to the specific entity that is the target of the person's actions
   b.       an actor's behavior in a situation is the same over time
   c.       other people in the same situation tend to respond like the actor
   d.       the cause of a behavior is internal or external

2.       The fundamental attribution error refers to
   a.       people's tendency to deal with someone else's behavior without trying to figure out what made them behave that way
   b.       people's tendency to go along with the majority opinion in deciding what caused an event rather than reasoning it out for themselves
   c.       people's tendency to ignore situational causes of behavior and favor internal explanations
   d.       people's tendency to ignore internal causes of behavior and favor external explanations

3. The scientific study of how people are influenced by social situations is called:
   a. group psychology
   b. social psychology
   c. sociobiology
   d. sociology

4. How does bad mood affect helping?
   a. it increases helping, thus making us feel worse
   b. it decreases empathy, thus making us feel worse
   c. it decreases helping, thus making us feel better
   d. it increases helping, thus making us feel better

5. Research on salience would suggest that a child's biosterous behavior will seem more annoying when:
   a. the child is in a group of other children
   b. the child is in a group of adults
   c. the child is alone
   d. the child has just come home from school and the parents are not home

6. The self-fulfilling prophecy is set into action by:
   a. expectations
   b. mental maps
   c. situational cues
   d. the false consensus effect

7. Your unique ideas about how a college class should be run, what a typical straight "A" student is like, and how a typical professor will act are all examples of:
   a. attitudes
   b. attributions
   c. schemas
   d. stereotypes

8. Which of the following is NOT listed as a factor that affects a communicator's ability to change attitudes?
   a. credibility
   b. physical attractiveness
   c. similarity
   d. trustworthiness

9. {T F} Attitudes involve emotions as well as cognitions

10. The degree to which an actor's behavior is rewarding or costly to the observer is called:
    a. actor-observer difference
    b. fundamental attribution error
    c. hedonic relevance
    d. salience of information

11. Persuasion is a form of communication in which one person tries to change the _____ of another.
   a.   attitudes
   b.   attributions
   c.   perceptions
   d.   personality

12. One possible reason for the similarity-attraction effect is that similar people:
   a.   don't take as long to get to know as dissimilar people
   b.   have nothing new to discuss
   c.   make each other feel vulnerable
   d.   validate one another's opinions

13. The _____ _____ effect is the tendency to overestimate the number of people who make the same inferences as we do.

14. The media is a powerful influence on our attitudes because it is frequently:
   a.   entertaining
   b.   our only source of information
   c.   rejecting
   d.   entertaining

15. When Dr. Smith reads a favorable student evaluation, she concludes that she is an excellent teacher.  When she reads an unfavorable one, she concludes that the student was unmotivated.  This illustrates which of the following?
   a.   actor-observer differences
   b.   egocentricity
   c.   salience
   d.   self-fulfilling prophecy

16. Given only the information below, which one of the following pairs of people would you predict are most likely to form a friendship?
   a.   Harry and Mike who live on different streets
   b.   Susan and Mary who live next door to each other in the same dorm
   c.   Jan and Marcia who live in different towns
   d.   Liz and Bob who live on different floors of the dorm

17. The ultimate aim of the attribution process is to:
   a.   blame others for our unacceptable behavior
   b.   explain other people's behavior
   c.   explain our reaction to other people's behavior
   d.   reinforce our own behavior

18. In recent election campaigns, there has been much criticism that campaign ads focus too much on building attractive images for candidates and too little on substantive issues of genuine importance.  These two ways of trying to influence voters are dealt with by the approach to persuasion called:
   a.   balance theory
   b.   dissonance theory
   c.   self-perception theory
   d.   the elaboration likelihood model

19.  In social cognitions, the primacy effect refers to
     a.   the most salient characteristic of a person
     b.   the tendency of our first impression of a person to bias our interpretation
          of their subsequent behavior
     c.   using the first person we meet in a new situation as a standard for
          judging the people we meet thereafter
     d.   Whatever trait we consider the most important for someone to have to
          function well in a specific situation

20.  Mindy is afraid of spiders.  In fact, Mindy is afraid of all insects.  Mindy's fear
     of spiders is a case where there is:
     a.   consensus
     b.   consistency
     c.   distinctiveness
     d.   relevance

21.  Research on physical attraction has shown that:
     a.   judgments of women's personalities are affected by their physical
          attractiveness, but judgments of men's personalities are not
     b.   men are more likely than women to make biased judgments of others
          based on physical appearance
     c.   most people disregard physical attractiveness when forming first
          impressions of people
     d.   we tend to ascribe desirable personality characteristics to good-looking
          people

22.  In dissonance theory, when we derogate attractive objects or activities that we
     forgo without strong external constraints, the situation is called:
     a.   effort justification
     b.   freely chosen justification
     c.   insufficient threat
     d.   just-barely-sufficient threat

23.  {T  F} The tendency to ignore personality, and thus to base one's judgments
          about another on observable situational constraints is called the
          fundamental consensus error.

24.  Our most strongly held and most difficult to change attitudes result from:
     a.   our personal experiences
     b.   parents
     c.   peers
     d.   the media

25.  According to Festinger, the feeling of discomfort that results from the
     realization that our beliefs and our behaviors are discrepant is called
     a.   attribution
     b.   balance theory
     c.   cognitive dissonance
     d.   confirmation bias

26. A major corporation believes that if its commercials simply present the true facts to potential buyers, it might be more likely to develop a long-lasting preference for their products. According to the elaboration likelihood model, this approach exemplifies the _____ route to persuasion
   a. autonomic
   b. central
   c. peripheral
   d. somatic

27. Research studying attitude change has focused on all of the following EXCEPT:
   a. the audience
   b. the communicator
   c. the context
   d. the message

28. The process of accepting a new position while giving up an old position is called
   _____ _____.

29. "Birds of a feather flock together." This statement supports which factor associated with attraction?
   a. differences
   b. proximity
   c. rewards
   d. similarity

30. If you expend much effort in reaching a goal:
   a. it is impossible to experience cognitive dissonance
   b. you will tend to evaluate the goal favorably in order to justify the effort
   c. you will conform to reach the goal no matter what the costs
   d. you will tend to evaluate the goal unfavorably, since it couldn't possibly have been worth the effort

31. Credible sources have the strongest effect on our attitudes:
   a. after a significant passage of time
   b. immediately after we hear the message
   c. through the peripheral route
   d. through the central route

32. {T F} People tend to attribute their successes to their own internal traits, while attributing their failures to situational factors.

33. A motivational state with the ultimate goal of increasing one's own welfare is called:
   a. altruism
   b. egoism
   c. egocentrism
   d. empathy

34. If you needed help in an emergency, in which of the following situations would you be most likely to get it?
   a. only one bystander, to whom there is some uncertainty if your situation really is an emergency
   b. only one bystander, who is sure your situation is an emergency
   c. several bystanders, all of whom are sure your situation is an emergency
   d. several bystanders, none of whom is certain if your situation is really an emergency

35. Greater emphasis on studying the relationship between people and groups during the 1960s and 1970s was influenced by:
   a. civil rights and Vietnam War
   b. Kurt Lewin
   c. Leon Festinger
   d. the Korean War

36. A belief that may have been false initially can sometimes lead to behaviors that make it come true. This is often called a(n)
   a. confirmation error
   b. self-fulfilling prophecy
   c. selection bias
   d. retrieval bias

37. Actors and observers tend to give different explanations for the same instance of behavior because:
   a. observers are less aware of the situational forces affecting actors' behavior
   b. observers tend to possess more knowledge about the actors
   c. only actors themselves can accurately explain their own behavior
   d. only outside observers can accurately explain an actor's behavior

38. Overemphasizing behavior and underemphasizing context refers to:
   a. failing to give sufficient weight to the situation in which a behavior occurred
   b. second guessing someone else's motives
   c. the degree to which one assigns a characteristic to another
   d. the difference between actor and observer attributions

39. The inferences that people draw about the causes of events, others' behavior, and their own behavior is called:
   a. attitudes
   b. attributions
   c. schemas
   d. stereotypes

40. Which of the following statements about the primacy effect is NOT true?
   a. Effects are strongest when we are judging stable characteristics.
   b. It has the strongest influence on the overall impression.
   c. It is the only influence on an overall impression.
   d. Your first encounter with a person should be your best.

41. In Kelley's attributional model, the dimension of consensus refers to whether:
    a. a person's behavior is unique to the specific entity that is the target of the person's actions
    b. an actor's behavior in a situation is the same over time
    c. other people in the same situation tend to respond like the actor
    d. the cause of a behavior is internal or external

42. Which of the following is an example of a dissonant relationship between a belief and behavior?
    a. an advocate for clean environment throws candy wrappings our the window while eating candy in a moving automobile
    b. a believer in physical fitness jogs every day
    c. a strong believer in birth control goes on a fishing trip over the weekend
    d. a strong believer in birth control volunteers to distribute pamphlets for the Planned Parenthood chapter

43. Rules that govern specific behavior and apply to all members of the group are called _____.

44. In the initial stages of dating, the variable that best predicts people's desire to go out with their date again is the date's:
    a. income
    b. intelligence
    c. personality
    d. physical attractiveness

45. {T  F} Although we are motivated to maintain consistency between our attitudes and our behavior, sometimes situational pressures force us to behave in attitude-discrepant ways.

46. Becky is having trouble working on a computer. When determining whether her trouble is due to an internal or external cause, which of the following questions best illustrates the element of consensus?
    a. Does Becky have trouble with all computers or only this one?
    b. Does Becky have trouble every time she uses this computer?
    c. Does Becky have the option of using other computers?
    d. Do others have trouble when working with this computer?

47. In Western cultures people tend to view the self as _____ while in Eastern cultures people tend to view the self as _____.
    a. independent; interdependent
    b. interdependent; independent
    c. inherently good; inherently bad
    d. inherently bad; inherently good

48. _____ place positive or negative meaning on the object or event while _____ are statements which express a relationship between objects or events.
    a. Attitudes; emotions
    b. Attitudes; beliefs
    c. Evaluation; norms
    d. Evaluation; beliefs

378

49. Chuck and Beth are both watching a commercial for a politician. Chuck decides that he will vote for the politician because she seems nice and trustworthy while Beth is persuaded to vote for the politician because she is convinced by her arguments. In this example, Chuck took the _____ route to persuasion while Beth took the _____ route.
    a. systematic; peripheral
    b. heuristic; systematic
    c. central; peripheral
    d. peripheral; systematic

50. Joe and Donna are close friends. Favors flow freely between them and they rarely keep track of how much they owe one another. Joe and Donna most likely have what type of relationship?
    a. exchange
    b. reciprocity
    c. homeostatic
    d. communal

## POSTTEST ANSWERS

1. b
2. c
3. b
4. d
5. b
6. a
7. c
8. b
9. T
10. c
11. a
12. d
13. false consensus
14. b
15. b
16. b
17. b
18. d
19. b
20. c
21. d
22. d
23. F
24. a
25. c
26. b
27. c
28. attitude change
29. d
30. b
31. b
32. T
33. b
34. b
35. a
36. b
37. a
38. a
39. b
40. c
41. c
42. a
43. norms
44. d
45. T
46. d
47. a
48. d
49. c
50. d

**PSYCH JOURNAL**

    **Please use the following pages to record your thoughts and feelings about the following questions.**

1.    How do you form first impressions? Have you ever judged someone unfairly or inaccurately based on a first impression? How long was it before you changed your opinion of that person? What made you change your mind?

_____

_____

_____

_____

_____

_____

_____

_____

_____

_____

_____

_____

_____

_____

_____

_____

_____

2. Think about the last time that you felt angry or irritated at someone. In describing the event to a friend, would you attribute your irritation to the person's character traits or to the particular situation in which the problem occurred?

_____

_____

_____

_____

_____

_____

_____

_____

3. What attitudes do you carry with you from home, whether inherited or learned from your parents? Have you found your attitudes changing as you are exposed to the world from the college perspective? Have you had any disagreements with your parents because your attitudes are changing?

_____

_____

_____

_____

_____

_____

_____

_____

_____

_____

4.  Have you ever worked really hard at a goal, achieved it, and then felt very proud of the accomplishment?  Does it mean more to you because it was difficult to accomplish?  Would you feel the same way if it had been easier?

_____

_____

_____

_____

_____

_____

_____

_____

5.  Describe a situation in which you had difficulty communicating with someone else because you didn't understand each other's culture, even though you (technically) spoke the same language?

_____

_____

_____

_____

_____

_____

_____

_____

_____

_____

6. Think about your friends. What attracted you to these people originally? What traits do you look for in friends? In a long-standing friendship, describe how your relationship has changed and evolved over time.

_____

_____

_____

_____

_____

_____

_____

_____

_____

_____

_____

_____

_____

_____

_____

_____

_____

_____

_____

_____

# Chapter 16 -- The Individual in Groups

## CHAPTER SUMMARY

1.     For the most part, social psychologists deal with small groups consisting of two or more people interacting in such a manner that each person influences the others. Organizational psychology expanded the study of groups to include work teams and large organizations that involve teams organized in networks. Cultural influences on groups are also studied.

2.     People belong to groups for many reasons. They may receive rewards. They may be seeking social support or information. Groups may also be useful in terms of self-evaluation, establishing self-identity, and achieving goals.

3.     Almost all groups develop norms, which are rules that apply to all members. Norms define what must be done and when. Roles are norms that apply only to people in certain positions. Roles define the obligations and expectations of those positions. Social dilemmas result when an individual's needs conflict with the needs of the group. Social dilemmas are difficult to resolve. Groups use norms and appeals to group loyalty to encourage members to act in the interest of the group. They also develop norms to punish individuals who act in their self-interest. Groups develop through predictable stages and group members' behavior changes from stage to stage.

4.     Conformity occurs when people change their attitudes or behaviors as a result of real or imagined group pressure, despite personal feelings to the contrary. Informational pressure causes people to match their attitudes and behavior to those of the group. Because belonging to groups is so important in everyone's life and because rejection and ridicule by the group are often painful, the group can exert normative social pressure to influence the behavior of its members. The size of the group, cultural factors, and the difficulty of the task are factors in conformity. Unlike conformity, obedience involves following the direct and explicit orders of a leader. Stanley Milgram showed that a surprising number of people would follow the orders of an authority figure even when they were ordered to hurt someone. Ideological zeal, personal gratification, and material gain, as well as a lack of responsibility, are some of the reasons people are willing to follow orders.

5.     There is general agreement that many of the behaviors and practices involved in leadership can be learned. Groups often have two leaders: a task specialist who is most involved with getting the job done and a socioemotional leader who reduces group tension and encourages members to contribute. The most current view finds that those who emerge as leaders are affected by the characteristics of the leader, followers, and situation. No single style of leadership is always effective. Different styles are effective in different situations. According to Conger, charismatic leaders are often agents of change with great energy who hold a vision of the future.

6.     Research on decision making by small groups of executives in large companies revealed three decision making processes: identifying the problem, developing a list of alternative solutions, and selecting a solution. People in a group tend to take stronger stands on issues than they would on their own. This group polarization effect may occur because the group rather than the individual will be held responsible for the decision -- the individual then feels freer to adopt an extreme position. It also seems that people in groups compare their attitudes and behaviors to

others in the group; to avoid looking undecided, each group member tries to adopt a position that is at least as strong as the positions of the other group members. When group members are more concerned about arriving at a group consensus than in making the best decision, groupthink occurs. A strong leader reinforces groupthink since group members may falsely assume that the leader cannot be wrong.

In order to avoid groupthink, group members must be encouraged to express their opinions even if they disagree with other group members. Outside information and opinions must also be sought. How computer-mediated groups reach decisions is not yet clear. Because people can suggest ideas anonymously, many new ideas are generated in these groups. However, there is an indication that these groups are less productive than face-to-face groups, especially when a task requires group consensus. A group member's status can determine the extent of the person's influence on the group. The Japanese encourage lower-echelon employees to initiate proposals in order to circumvent this "status rules" factor. A minority's best chance for influencing decision making is to be clear, consistent, and self-confident.

7.      Social facilitation occurs because the presence of other people tends to arouse people; this arousal creates an additional pool of energy that aids the performance of well-learned behaviors. Social loafing, or less than full effort, may result if people perform tasks as part of a group effort and if their individual input on the task is not readily identifiable.

8.      There are a number of theories about why prejudice and discrimination develop: one places the blame on the formation of groups creating "ingroups" and "outgroups" often in competition with each other; another suggests that people learn how to be prejudiced from parents and the media; a third argues that prejudice and discrimination are the result of scapegoating. Reducing prejudice and discrimination are very difficult. People need to make equal status contact. Prejudice is reduced when people are placed in situations where they must work with the people or groups they are prejudice against to achieve a common goal. Creating situations in which children who have never suffered discrimination feel prejudice themselves has resulted in those children being less likely to develop prejudice and discrimination towards others. Traditional attempts to reduce prejudice and discrimination against African Americans are based on asking African Americans to assimilate -- to give up their cultural and racial identity. This is an unfair request. We might look to other countries, such as New Zealand, that support a multicultural approach to racial groups for solutions to prejudice and discrimination in the United States.


**KEY TERMS AND CONCEPTS**

The History and Scope of Studying Individuals in Groups
        Social psychologists deal with groups of two or more
        Organizational psychology expanded the study of groups
                Includes work teams and large organizations
        Cultural influences on groups is also studied

Why Belong to a Group?
        Reasons for belonging to groups
                Rewards are received
                Seeking social support or information
                Useful in terms of self-evaluation, and establishing self-identity
                Achieving goals

The Structure of Groups
  Norms - rules that apply to all members of a group
    Written and unwritten (codes)
    Define what must be done and when
  Roles - rules that apply only to persons in certain positions
    Define the obligations and expectations of those positions
    Influence how observers interpret behavior
    May severely limit the freedom of group members
  Norms and roles
    Together they give the group structure
    They can also prove disruptive and troublesome
    People belong to many groups, a role conflict may result
      Demands of one group may conflict with those of another
  Social dilemmas: groups versus individuals
    Result when an individual's needs conflict with needs of the group
    Norms and appeals to group loyalty are used to encourage members
    Norms punish individuals who act in their self-interest
    Difficult to resolve
  Group change and development
    Groups develop through predictable stages
      Forming, storming, norming, performing, and adjourning
    Groups change from stage to stage

Conformity: Dancing to the Group's Tune
  Why we conform
    We conform when people change attitudes as a result of group pressure
      This happens despite personal feelings to the contrary
    The group possesses normative social pressure
      Belonging to groups is so important in everyone's life
      Rejection and ridicule by the group are often painful
    Simple compliance - change in behavior but not private attitude
    Private acceptance - change in both public and private attitude
  Factors influencing conformity
    The individual
    Size of the group - the bigger the better
    The task - how difficult it is will influence conformity
  Obedience: "I was only following orders"
    Following the direct and explicit orders of a leader
    Stanley Milgram's study on obedience
      Subjects followed orders even when they were hurting someone
    Reasons people follow orders
      Ideological zeal, personal gratification, and material gain

Leadership
  Everyone is both a leader and a follower, some may lead more than others
  Theories of leadership
    Many of the behaviors and practices involved in leadership are learned
    There are often two types of leaders in a group
      Task specialist - getting the job done
      Socioemotional - reduces group tension
    Those who emerge as leaders are affected by three things
      Combination of leader traits
      The followers
      Situation

The charismatic leader: the magical force
Agents of change with great energy and a vision for the future

Group Decision Making
Research revealed three decision making processes
Identifying the problem
Developing a list of alternative solutions
Selecting a solution
Group polarization
People in a group tend to take stronger stands on issues
Occurs because the group would be held responsible not the individual
The individual then feels freer to adopt extreme positions
People compare their attitudes to others in the group
Social comparison approach
Those in the group want to avoid looking undecided
Persuasive argument hypothesis - why people hold certain positions
Evident when a group is faced with intellectual or factual issues
Groupthink
Occurs when group members are more interested in arriving at consensus
A strong leader reinforces groupthink
Groupthink occurs for three reasons
Members are strongly attracted to a group
Members falsely assume the leader cannot be wrong
There is a large group
Avoiding groupthink
Group members must be encouraged to express their opinions
Even if they disagree with other members
Leader should refrain from stating a preference
Outside information and opinions must also be sought
Several meetings should be called to reassess new information
Groups can be subdivided into smaller groups
Technology and group process
Not clear how computer-mediated groups reach decisions
Anonymous ideas generate many new ideas
These groups are less productive than face-to-face groups
This is especially true when a task requires group consensus
Who influences the decision
A group member's status can determine the extent of influence

Group Performance
General types of problems that plague groups
Coordination loss
Motivation loss
Social facilitation: lending a helping hand
Presence of others tends to arouse people
This creates an additional pool of energy that aids performance
Social Loafing: A place to hide
Results if people perform tasks as part of a group effort
Individual input is not readily identifiable

Intergroup Relations, Prejudice, and Discrimination
Prejudice - unjustified negative attitude toward an individual
Stereotypes - oversimplified generalizations about the characteristics of a group
Discrimination - aggressive behavior aimed at the target of prejudice

Exclusion from social clubs, or jobs are examples
The roots of prejudice and intergroup conflict
    Group categorization
        Formation of "ingroups" and "outgroups" in competition
    Competition
        Two or more parties attempt to reach a goal
        Conflict and hatred often escalate
    Learning
        Learn from parents and media
        Difficult to change if supported by peers
    Scapegoat theory
        Minority group is singled out as the target of aggression
           Safe to attack and highly visible characteristics
Reducing prejudice and discrimination - very difficult
    Equal-status contact
        Increase contact between members of different groups
    Cooperation
        People are placed with others to work towards common goals
    Experiencing prejudice
        Children who never suffered discrimination feel prejudice
           These children are less likely to develop prejudice
Reducing racial prejudice: a penetrating reexamination
    Traditional attempts to reduce prejudice and discrimination
        Asking African Americans to give up culture and racial identity
           This is an unfair request
    Other countries set an example for solutions to prejudice in the U.S.

## DISCUSSION QUESTIONS AND EXERCISES

**Note: These questions and exercises ARE the learning objectives for this chapter. Answer them accurately in your own words and you will have mastered the most important material. We guarantee it.**

1.    The History and Scope of Studying Individuals in Groups

a.    Discuss each of the following group's approach and perspective:

    (1)    Social psychologists

    (2)    Organizational psychology

(3)    Cross-cultural behavior

2.    <u>Why</u> <u>Belong</u> <u>to</u> <u>a</u> <u>Group?</u>

a.    Identify and describe at least four reasons we join groups.

3.    <u>The</u> <u>Structure</u> <u>of</u> <u>Groups</u>

a.    What are **norms**? What do they specify?

b.    What are **roles** and what do they define? What do norms and roles do together? What are some of the problems that occur with norms and roles?

c.    What is the **social dilemma**?  Why do individuals in groups often act out of self-interest?

d.    Why are social dilemmas difficult to resolve?

e.    Identify the predictable stages of group development.  Give examples of three of the five stages.

f.    What are some of the characteristics of a group that were found in research, when a group is just forming and through the rest of the stages?

4. Conformity: Dancing to the Group's Tune

a. How would you define **conformity** in your own words? Can conformity be dangerous? Why or why not?

b. Discuss some of the reasons why we conform. Have you ever experienced any of these reasons to conform to a group?

c. What is the difference between simple compliance and private acceptance?

d. Discuss each of the following factors that influence conformity.

   (1) The individual

(2)     The group

(3)     The task

e.      How does **obedience** differ from conformity?  Discuss some of the findings
        from Stanley Milgram's experiment.

f.      What do these studies of obedience suggest?  Identify and describe the other
        factors that underlie our willingness to follow orders.

g.      How does our role influence whether or not we will follow orders?

5. <u>Leadership</u>

a. How is a leader defined? What are some observations that this definition has given rise to?

b. How is a task specialist different from a socioemotional leader?

c. Discuss some of the various theories of leadership. Is there a magic trait?

d. What is charisma? How do we get it?

6.    <u>Group</u> <u>Decision</u> <u>Making</u>

a.    Research on decision making by small groups of executives in large companies showed that actions were really based on three decisions.  What are they?

b.    Describe the **group polarization effect**.  Where does it occur and what are some of the reasons for this effect?

c.    Why do we want to avoid **groupthink**?  When is it most likely to occur?

d.    Identify the methods that Janis suggests for avoiding groupthink.

e.      What role has technology played in the group process?  What are the pros and cons of computer-mediated groups?

f.      How is the power to influence decisions found?  How might cultures differ in this area?  Give some examples.

7.    Group Performance

a.      What are some of the factors to determine whether an individual is more effective than a group in performing a certain task?

b.      What is **social facilitation**?  What have studies shown with regard to the presence of an audience and its effects on our performance?

c.  What is **social loafing**?  The social facilitation and social loafing research seem to contradict each other.  Explain the difference between these two bodies of research.

8.  <u>Intergroup</u> Relations, Prejudice, and Discrimination

a.  What is the difference between **prejudice** and **discrimination**?  What are **stereotypes**?

b.  What are some of the characteristics of prejudice and discrimination that make them more difficult to understand than any other behavior or attitude?

c.  Describe how each of the following situational factors may influence prejudice:

    (1)  Group categorization

    (2)  Competition

(3)     Learning


(4)     Scapegoat theory




d.     Identify and describe some of the methods used to reduce prejudice and
       discrimination.




**POSTTEST**

1.     Social loafing refers to:
       a.     a tendency to blame others for the group's poor performance
       b.     increases in socializing among members of larger groups
       c.     the loss of coordination among group members' efforts
       d.     the reduction of effort by individuals when they work in groups

2.     The subfield of psychology most interested in how groups affect job
       performance and adjustment is known as _____ psychology.
       a.     community
       b.     industrial
       c.     organizational
       d.     social

3.     {T F} Sam wants to be part of the popular crowd at school. When the popular
       kids all get earrings, Sam, not wanting to be ridiculed, also gets an
       earring. In this example, the group has the influence of normative social
       pressure.

4.     Of the following reasons for joining a group, which one is MOST similar to
       rewards?
       a.     achieving a goal
       b.     gaining social support
       c.     information and evaluation
       d.     obtaining a self identity

5.    The idea that we often learn about ourselves by comparing our performance
      with that of other people is reflective of:
      a.    cognitive dissonance
      b.    performance comparison theory
      c.    social comparison theory
      d.    social reality

6.    Of the following, which BEST explains why many people join self-help groups?
      a.    achieving a goal
      b.    rewards
      c.    to fit in
      d.    to gain social support

7.    Prejudice:
      a.    is the same thing as discrimination
      b.    refers to a negative attitude toward members of a group
      c.    refers to unfair behavior toward the members of a group
      d.    both a and b

8.    {T  F} Standing when the national anthem is played is an example of a norm.

9.    In Milgram's (1963) study of obedience, subjects:
      a.    became the recipients of painful electric shocks delivered by an
            experimental accomplice
      b.    indicated which of three lines matched a "standard line" in length
      c.    were ordered to deliver painful electric shocks to a stranger
      d.    were ordered to give consistently wrong answers to simple questions

10.   Which of the following is NOT an example of a role?
      a.    daughter
      b.    human
      c.    nurse
      d.    parent

11.   The beliefs, attitudes, and behavior of other people are referred to as

      _____ _____.

12.   In Asch's studies, what was the relationship between group size and
      conformity?
      a.    Conformity did not change with group size.
      b.    Conformity increased as group size went from two to four, and then
            decreased.
      c.    Conformity increased steadily as group size was increased up to fifteen.
      d.    Conformity increased as group size increased up to four members, but
            after that point, it did not lead to higher conformity

13.   Howie is a member of a rock band and a professor.  To be accepted in the band
      and by fans, Howie must wear his hair long.  However, to be accepted in the
      education field, he must wear his hair short.  This is called:
      a.    avoidance-avoidance conflict
      b.    conflict of norms
      c.    role conflict
      d.    social dilemma

14. According to social psychologists, a group:
    a. consists of two or more people who interact and are interdependent
    b. consists of three or more people who interact on a regular basis
    c. exists whenever two or more people are in spatial proximity to each other
    d. will not affect the behavior of its members

15. What happens when a person's needs are in conflict with the groups needs?
    a. a social dilemma results
    b. deindividuation
    c. group polarization develops
    d. reactance to the conflict

16. {T F} Early research efforts focused on the great person leadership theory.

17. Research has found that groups often go through predictable stages. Which stages is the most difficult for members?
    a. adjourning
    b. forming
    c. norming
    d. performing

18. Hillary really liked to play with her next door neighbor, but she did not play with her because of normative social pressure, even though she wanted to. What concept does this illustrate?
    a. obedience
    b. private acceptance
    c. reactance
    d. simple compliance

19. Discrimination:
    a. is the same thing as prejudice
    b. is oversimplified generalizations about the characteristics of a group
    c. refers to a negative attitude toward members of a group
    d. refers to unfair behavior toward the members of a group

20. Early studies indicated that women conformed more than men. Later research has shown that women's conforming behavior is motivated by their desire to:
    a. avoid rejection
    b. be popular
    c. become the group leader
    d. keep group conflict at a minimum

21. {T F} If you were to experience a change in your public behavior and private attitudes it would be referred to as simple compliance.

22. Which of the following are most likely to conform to the norms of the sorority?
    a. alumni
    b. graduating seniors
    c. officers
    d. pledges

23. A group is considered to have _____ pressure, when the group has influential value as a source of information.

24.  {T  F} Conformity occurs when people change their behavior after observing a model being reinforced for a particular response.

25.  In the Hofling (1966) study on obedience, the nurses:
     a.  followed orders over the phone from a doctor they did not know
     b.  refused to double a dosage because they did not know the doctor
     c.  reported the phone call to a hospital administrator
     d.  scared patients with the orders of an unknown doctor

26.  Obedience is a form of compliance in which people change their behavior in response to:
     a.  direct commands
     b.  implied pressure
     c.  persuasive communications
     d.  requests from others

27.  Research has shown that the person who _____ often becomes the group's leader.
     a.  arrives at the meeting first
     b.  is seated at the head of the table
     c.  possesses certain traits that would go with leadership
     d.  remains standing when everyone else is sitting

28.  {T  F} Charismatic leaders often pay close attention to the details.

29.  Jodie and Matt are the co-captains of their tennis team.  While the coach decides who plays in each match, Jodie and Matt have the responsibility of maintaining the team's morale.  What type of leaders are Jodie and Matt?
     a.  great person
     b.  relationship-motivators
     c.  socioemotional
     d.  task specialist

30.  Karen was unsure of her position on the environment, but she was leaning in favor of it.  At the health club, she found herself defending the issue.  This illustrates:
     a.  deindividuation
     b.  group cohesiveness
     c.  group polarization
     d.  groupthink

31.  What is the most important feature of minority influence?
     a.  clear position
     b.  consistency
     c.  ingratiation
     d.  self-confidence

32. According to the persuasive argument hypothesis, the group polarization effect occurs because:
   a. individuals find that their original position is not as extreme as the position of others
   b. individuals do not feel responsible for a group decision
   c. individuals compare the views of others in the group to theirs
   d. it is the information about why people hold certain positions that is most important

33. Groupthink occurs when members of a cohesive group:
   a. are initially unanimous about an issue
   b. emphasize concurrence at the expense of critical thinking in arriving at a decision
   c. shift toward a less extreme position after group discussion
   d. stress the importance of caution in group decision making

34. The major problem associated with group think is which of the following?
   a. group polarization occurs
   b. no critical evaluation is done
   c. risky shift occurs
   d. there is no leadership during the decision making process

35. The results of Milgram's (1963) study imply that:
   a. in the real world, most people will refuse to follow orders to inflict harm on a stranger
   b. many people will obey an authority figure even if innocent people get hurt
   c. most people are willing to give obviously wrong answers when ordered to do so
   d. most people stick to their own judgment, even when group members unanimously disagree

36. Of the following, which is one of the ways to avoid groupthink?
   a. allow outsiders to attend meetings and to evaluate decisions
   b. do not consult outsiders until a decision is made
   c. have all group members suspend their judgments on decisions
   d. have the members and the leader state their preferred solutions

37. Asch found that group size made little difference if:
   a. just one person fails to go along with the rest of the group
   b. the experimenter ridicules the group's wrong answers
   c. the task was easy
   d. the task was difficult

38. The presence of others leads to social _____ of well-learned responses.

39. {T F} Roles and norms, together, give a group structure.

40. When group members feel their efforts are going unrecognized,
_____ occurs.
a. coordination loss
b. groupthink
c. inconsistency
d. motivation loss

41. _____ puts people into categories with each category having its own set of characteristics.
a. bias
b. discrimination
c. prejudice
d. stereotyping

42. The group polarization effect implies that:
a. group decisions will always be better than individual decisions
b. the gap between two opposing factions will be narrowed after group discussion
c. when most of the group members initially favor a cautious decision, discussion will cause the group to adopt a risky decision
d. when most of the group members initially favor a cautious decision, discussion will cause the group to adopt an even more cautious decision

43. According to the scapegoat theory, prejudice and discrimination result from:
a. imitation
b. frustration
c. religion
d. reinforcement

44. Which of the following statements is true?
a. Derogatory stereotypes no longer exit today.
b. Ethnic and racial groups are the only targets of widespread prejudice.
c. People tend to see what they expect to see when they come into contact with members of those groups that they view with prejudice.
d. People tend to selectively recall instances that counteract their stereotypes.

45. According to research, prejudice is reduced when people must:
a. change their attitudes and behaviors
b. come into contact with one another
c. cooperate with one another to achieve a common goal
d. live together

46. The teaching faculty at a new college are currently trying to decide if they will make their decision by a majority vote or by group consensus. This group is most likely in what stage of development?
a. norming
b. performing
c. storming
d. forming

47. The emergence of a successful leader is determined by leader traits, group needs, and the situation. This statement best reflects which approach to leadership?
   a. interaction
   b. situational
   c. environmental
   d. great person

48. Research on decision making by small groups of executives in large companies showed that actions were based on all of the following except:
   a. identify the problem
   b. develop a list of alternative solutions
   c. gather appropriate information
   d. select a solution

49. A group of five managers were going to meet to decide how much money they should spend on office equipment. Before the meeting, each of the managers wanted to purchase moderately expensive equipment. Following the meeting, however, the group decided to purchase the most expensive equipment available. This is an example of the
   a. false consensus effect
   b. pluralistic ignorance
   c. induced compliance
   d. group polarization effect

50. Suppose you were playing on a volleyball team against a long-time rival and the opposing team makes an awful play. Members of your team would most likely view that awful play to be a result of:
   a. the opposing team's poor ability
   b. the opposing team's bad luck
   c. bad weather conditions
   d. the opposing team's lack of effort

**POSTTEST ANSWERS**

1.     d
2.     c
3.     T
4.     a
5.     c
6.     a
7.     b
8.     T
9.     c
10.     b
11.     social reality
12.     d
13.     c
14.     a
15.     a
16.     T
17.     b
18.     d
19.     d
20.     d
21.     F
22.     d
23.     informational
24.     F
25.     a
26.     a
27.     b
28.     F
29.     c
30.     c
31.     b
32.     d
33.     b
34.     b
35.     b
36.     a
37.     a
38.     facilitation
39.     T
40.     d
41.     d
42.     d
43.     b
44.     c
45.     c
46.     a
47.     a
48.     c
49.     d
50.     a

# PSYCH JOURNAL

**Please use the following pages to record your thoughts and feelings about the following questions.**

1.  Certain groups like the Crawfords seem to work smoothly, and function well together. Based on what you have learned about group behavior, identify the factors that distinguish between groups that work and those that do not. How would you go about setting up a group that was effective?

_____

_____

_____

_____

_____

_____

_____

_____

_____

_____

_____

_____

_____

_____

_____

_____

_____

2.   List the groups to which you belong.  Which are the most important and influential to you?  Do you behave differently in some of these groups than you do by yourself?

_____

_____

_____

_____

_____

_____

_____

_____

_____

_____

_____

_____

_____

_____

_____

_____

_____

_____

_____

_____

3.  How have some of the groups in which you belong changed over time?  Do you feel differently about belonging to those groups than you did originally?

_____

_____

_____

_____

_____

_____

_____

_____

4.  Have you ever had to choose between what was good for you as an individual versus what was good for the group?  How did you choose?

_____

_____

_____

_____

_____

_____

_____

_____

_____

_____

_____

_____

5. Think about the group projects you have been assigned in the past. Which were the best experiences? The worst? Explain what you think makes for the best possible group assignment.

_____

_____

_____

_____

_____

_____

_____

_____

6. Have you ever failed to speak up when you thought a group decision was a poor one because you didn't want to "make waves"? Explain what happened.

_____

_____

_____

_____

_____

_____

_____

_____

_____

_____

_____

_____

7. Do you hold any strong feelings about particular groups of people, even if you have never met anyone from that group? How did your attitude form? What would it take to change your mind?

_____

_____

_____

_____

_____

_____

_____

_____

_____

_____

_____

_____

_____

_____

_____

_____

_____

_____

_____

_____